TABLE

its symbol and the atomic number below its symbol.

									4.003
									He
									2
			10.81	12.011	14.007	16.000	19.00	20.182
			B	C	N	O	F	Ne
			5	6	7	8	9	10
			26.98	28.09	30.97	32.066	35.453	39.946
			Al	Si	P	S	Cl	A
			13	14	15	16	17	18
58.71	63.54	65.38	69.72	72.59	74.91	78.96	79.91	83.80
Ni	Cu	Zn	Ga	Ge	As	Se	Br	Kr
28	29	30	31	32	33	34	35	36
106.4	107.87	112.40	114.82	118.69	121.74	127.61	126.91	131.3
Pd	Ag	Cd	In	Sn	Sb	Te	I	Xe
46	47	48	49	50	51	52	53	54
195.09	197.0	200.6	204.37	207.2	209.0	210	211	222
Pt	Au	Hg	Tl	Pb	Bi	Po	At	Rn
78	79	80	81	82	83	84	85	86

157.25	158.92	162.50	164.93	167.3	169	173.04	174.98
Gd	Tb	Dy	Ho	Er	Tm	Yb	Lu
64	65	66	67	68	69	70	71
247	249	251	254	255	256	255	257
Cm	Bk	Cf	E	Fm	Mv	(No)	Lw
96	97	98	99	100	101	102	103

GEOCHEMISTRY
OF
SOLIDS

An Introduction

GEOCHEMISTRY
OF
SOLIDS

An Introduction

W. S. FYFE
University of California, Berkeley

McGRAW-HILL BOOK COMPANY

New York · San Francisco · Toronto · London

Geochemistry of Solids: An Introduction

Copyright © 1964 by McGraw-Hill, Inc. All Rights Reserved.
Printed in the United States of America. This book, or parts thereof,
may not be reproduced in any form without permission of the pub-
lishers. *Library of Congress Catalog Card Number* 63-17338

22645

23456789 HDBP 10987

PREFACE

This book is addressed to students of the earth sciences who are studying elementary mineralogy, petrology, and geochemistry, and to students of inorganic chemistry who have an interest in the solid state.

The increase in the number of workers in every field of science, and the associated increase in the amount of factual data, often has the undesirable consequence that students of the more descriptive sciences are given less time to study the three primary disciplines of physical science: mathematics, physics, and chemistry. Thus students of the earth sciences, both at elementary and advanced levels, frequently have only the most rudimentary ideas of such things as atomic structure and chemical bonding. But the complex problems which confront the student of natural science make this fundamental background quite essential.

This book, it is hoped, may be used to supplement some of the standard and larger works and may open the way to greater understanding of what are, to this writer at least, fascinating aspects of our studies. The book is not comprehensive and is not meant to be. It will be successful if it leads some to follow on with the more advanced texts listed in the final brief bibliography. Those who read this book may frequently wish to refer to one of the more modern standard texts of mineralogy or structural inorganic chemistry.

The writer gratefully acknowledges the help of those who have read this manuscript, particularly Professors C. O. Hutton, Brian Mason, and John Verhoogen. But all inadequacies and errors are the writer's. Finally, thanks are due to those structural chemists and mineralogists whose illustrations of crystal structures have been frequently copied. These data form, in large part, the core of our science. In particular the writer thanks Cornell University Press and

Sir Lawrence Bragg for permission to use some illustrations from Bragg, "Atomic Structure of Minerals." Several illustrations concerning dislocations were reproduced from Read, "Dislocations in Crystals," McGraw-Hill Book Company, Inc., New York, 1953.

W. S. Fyfe

CONTENTS

Chapter 1

THE SOLID STATE

Mineralogy, petrology, and crystal chemistry are concerned with the arrangement and behavior of atoms, molecules, and ions in solids. At the outset it is perhaps necessary to clarify what is meant by the term *solid* and how this state of matter is to be distinguished from other states or arrangements of atoms and molecules. Classification must involve certain criteria, and, as with all rigorous attempts to classify, there are some states that will not fit. Classification must thus be considered primarily as a means of drawing together certain materials with much in common to set convenient limits on the scope of our discussion.

Three major states of a given system (a system being that part of the universe on which we have chosen to fix our attention) are generally recognized: gaseous, liquid, and solid. If asked to explain to someone less experienced than ourselves what is meant by these terms, we would probably make the following types of statement. First, the gaseous state is one where the atoms and molecules are free to move around and have what may be called *freedom of translation*, and the structure or molecular arrangement is chaotic. This does not imply that there is no association or grouping of molecules as at any given instant there may well be, but these groups will typically be small and frequently forming and breaking apart. The one feature that must enter into our description, however, is that the gas has no unique volume and no boundaries, except those imposed upon the gas by the container we choose to place it in. This feature will serve to define a gas.

If we could inspect the solid from atomic dimensions, we would note a very large degree of order or regularity in the arrangement. In any fixed direction, we would pass the same features at regular intervals and see very few sites of irregularity. The solid thus exhibits what is called *long-range order* with the same features

1

repeated identically over very many molecular diameters. That such internal order is present is almost intuitively suggested by the surfaces of many solids. In the case of the true solid formed by most natural processes, we would find this boundary surface to be planar, perhaps broken by steps, but still for the greater part consisting of planar units. Thus the solid can be defined as any material which can occur spontaneously in a form bounded by plane surfaces, the exterior form reflecting the long-range order to be found in the interior.

The liquid state is in many ways intermediate between the two states already considered. The atoms and molecules of the liquid exhibit much greater order than those in the gas, but this order is short-range, persisting over a few molecular diameters only. Many molecules and groups of molecules would still be moving from position to position. Under suitable conditions the liquid has a boundary surface and this boundary surface may be curved, particularly if we form small drops of a liquid. The presence of a boundary indicates some definite volume; above this boundary, matter will be in the form of a gas. It should be noted that when the liquid coexists with its vapor at a given temperature, the number of atoms or molecules per unit volume of the liquid, or its density, is unique, a situation which does not occur with a gas. Let us place some liquid, e.g., water, in a glass container with a piston, and with the liquid half-filling the container. The space above the liquid contains water molecules, and as long as the temperature is constant and iquid is present, the pressure and density of the gas will be constant. Thus the liquid has a definite vapor pressure which depends on the temperature. If we withdraw the piston, more liquid will vaporize; if we advance the piston, more liquid will form by condensation. What happens if we push in the piston until the last portion of gas has vanished and the pressure on the piston is slightly in excess of the vapor pressure? Do we now have a liquid or gas? In fact, there is some ambiguity. We can now consider the system to be a gas or a compressed liquid. We can only justify the latter term by showing that on withdrawal of the piston, a curved boundary reforms. This will never happen when a gas is expanded at constant temperature. Such information can be usefully expressed in what is known as a *phase diagram*, which describes the state of a system of given chemical composition in relation to the temperature and pressure acting on it. In Fig. 1-1 such a diagram is shown for the system containing hydrogen and oxygen atoms in the ratio 2:1, the system water.

Inspection of this diagram indicates that at sufficiently low pressures at all temperatures above absolute zero, we can fill our container with a gas. At low temperatures and higher pressures the gas may condense to a solid form of water, which may occur in some forms well above 100°C if the pressure is sufficiently large.

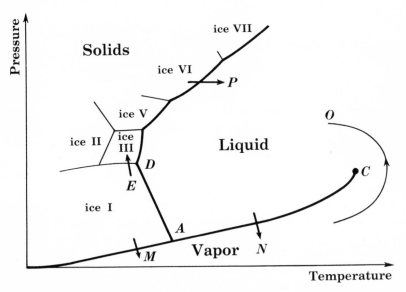

Fig. 1-1 Schematic representation of the various forms of water as a function of temperature and pressure. Ice I, II, etc., are solid forms (polymorphs) with different arrangement of water molecules. The point A, where liquid, vapor, and ice I coexist, occurs at a pressure of 0.006 atm and a temperature of 0.0099°C (in the absence of the atmosphere). Point D, where normal ice coexists with ice III and liquid, is at −22°C and 2047 atm pressure. Above this pressure, all ice forms are more dense than the liquid. The melting temperature of ice VII is 190°C at about 40,000 atm. Point C, the critical temperature of liquid water, where liquid and vapor states become indistinguishable, occurs at 374.15°C and a pressure of 218 atm. The volume of the liquid at this point is 3.08 cm³/g. For explanation of arrows, see text.

At very high temperatures the water dissociates into H_2 and O_2 molecules and eventually, as the temperature becomes extreme, into H and O atoms. Along the curve AC, it is possible to continually observe the boundary between liquid and vapor states at temperatures well above the familiar atmospheric boiling point of 100°C. Near 373°C the liquid commences to expand rapidly; and quite suddenly, as the temperature is raised further, the boundary vanishes.

The temperature at which this phenomenon occurs is called the *critical temperature*, and above this temperature the coexistence of liquid and vapor will not be observed. We can take a gas along the path O (Fig. 1-1) and pass from gas to compressed liquid without any visible change in the system. The term *fluid* is sometimes used to discuss states above the critical temperature, but the term is unnecessary and introduces the concept of flow which occurs in all forms of matter.

It is often erroneously implied that the properties of water change rather drastically as the critical region is passed. The solvent properties of water, for example, change quite smoothly through the critical region if the density is allowed to vary in a smooth fashion.

The three states we have briefly mentioned can be reasonably well defined, but other states do occur and sometimes cause confusion. First, when certain inorganic liquids are cooled very rapidly (e.g., most silicate melts), the stable solid state does not form, but we produce a very viscous liquid or *glass*. This glass may be similar in many ways to a solid except that it will again be bounded by curved surfaces and must be considered a true liquid lacking long-range order. There is also a class of compounds which produce so-called *liquid crystals* or occur in the *paracrystalline* state. Normally this state arises when a solid containing long rigid molecules commences to melt. The disorder induced by melting is often only partial, and groups of well-oriented molecules remain, imparting certain properties to the material associated with both solid and liquid states. The situation is indicated in Fig. 1-2. Finally there is the so-called *amorphous* state, not uncommon among minerals and chemical compounds. If a solution of sodium silicate is rapidly acidified, a gelatinous precipitate of hydrated silica is formed. This material if dried does not produce the reactions typical of a solid. We can probably consider such amorphous materials to consist of extremely small solid units, but the number of atoms per unit is so small that the forces which lead to the planar surfaces of solids and their internal order are distorted by the enormous number of atoms in surface positions. The state is one of great instability, and given suitable conditions, the small units will join to form the true solid.

Inspection of Fig. 1-1 indicates that we can pass directly from any form of matter to another, for example along path M, from solid to gas or gas to solid; along N, from liquid to gas or gas to liquid; along E, from solid to solid; along P, from solid to liquid; and along O, from gas to compressed liquid.   All these transitions are frequently involved in the formation of chemical substances and materials within the earth.

Most of our remaining discussion will thus be concerned with consideration of the properties óf solids.   There are a number of questions which must be considered:

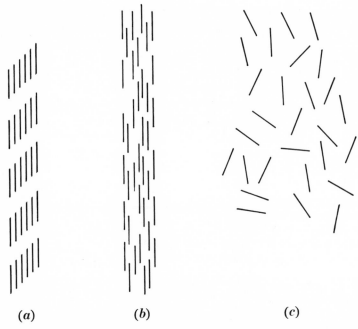

(a)   (b)   (c)

Fig. 1-2 Schematic illustration of three states of matter: (a) a solid with long-range order of the unit parts; (b) the paracrystalline state with partial order; and (c) the liquid or gas state with no long-range order but molecular chaos.

1. Why are these materials bounded by plane surfaces?

2. Why are the structures so regular in three dimensions, and what forces are responsible for the type of regularities observed?

3. What determines the common physical and chemical properties of these substances?

To answer such questions, we must first consider some general features of atomic structure and atomic interactions, for, ultimately, we would like to be able to discuss or predict these properties from the fundamental properties of the atoms in the solid.

A Digression on Energy and Stability

We are all familiar with systems in which an energy function is associated with stability or with tendency to change. As a familiar example, consider the simple case below (Fig. 1-3) where two identical blocks are situated on a flat table in two different positions.

If the center of gravity of the blocks is at g, we know that the most stable position occurs when this center of gravity is at the lowest possible elevation. In this situation, the gravitational potential

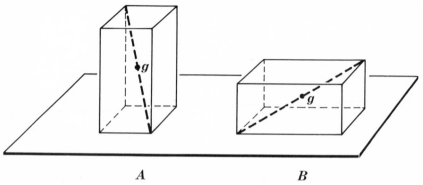

A   *B*

Fig. 1-3 Illustration of two stable situations in a gravitational field. Situation *A* is less stable than *B*, or is metastable with respect to *B* (see text).

energy is at a minimum. Both the blocks in Fig. 1-3 are quite stable, but position *B* is more stable than position *A*, or we could say that *A* is *metastable* with respect to *B*. A stable state is one which will persist indefinitely. In this case knowledge of the height of the center of gravity enables us to estimate the gravitational energy, and we can make definite statements about the relative stability of different positions. In making these statements, we are concerned with differences in energy and not absolute values.

In this book we are concerned with chemical systems containing a large number of atoms, and there are analogies to the mechanical systems above in that some states are more stable than others, and given suitable opportunities, a less stable state may change to a more stable or the most stable state. It might be anticipated that there is some function measuring "chemical energy" which would allow us to predict relative stabilities.

Consider a reaction A → B which is known to occur or which feasibly could occur. The compounds in state A and state B are known, and the equation is stoichiometrical or balanced. It is possible to estimate the energy of reaction ΔE and the heat of reaction ΔH. These could be directly measured if states A and B were available. ΔE for example, would represent the difference in internal energies of states B and A; that is,

$$\Delta E = E_\mathrm{B} - E_\mathrm{A}$$

We might guess that the heat absorbed or evolved would be a good guide as to relative stability, and, in fact, this is frequently the case. But application of the laws of thermodynamics, verified experimentally innumerable times, demonstrates that these functions alone are insufficient. To predict the relative stability and directions of change of states of a chemical system at constant pressure and temperature, the required function is called the *free energy* and designated G or F. This G is our measure of "chemical potential energy." In the process $A \rightarrow B$, if $\Delta G = G_\mathrm{B} - G_\mathrm{A}$ is negative, state B is more stable than A, and the reaction may proceed spontaneously; if ΔG is positive, our reaction will not proceed spontaneously; and if ΔG is zero, the two states are equally stable or are in chemical equilibrium.

The function ΔG is a compound function capable of direct measurement by several methods.

$$\Delta G = \Delta H - T\,\Delta S = \Delta E + P\,\Delta V - T\,\Delta S$$

where ΔE = change in internal energy
     ΔH = heat of reaction
       T = absolute temperature
       P = pressure
     ΔV = change in volume
     ΔS = change in entropy

It will be noticed that the final term involves the change in entropy. The concept of entropy may be less familiar, but in many ways entropy may be more readily envisaged than heat or energy. Like heat and energy, entropy is an extensive property of a chemical system, its value depends on the quantity of material, and its value is a unique function of pressure and temperature and physical state. The value is measured in calories per gram molecule per degree absolute, and hence TS has units of energy. Let us examine a few values of changes in this quantity to gain some insight into what entropy measures. At 0°C ice, liquid water, and vapor coexist.

For the various possible changes we have:

$$\text{Ice} \rightarrow \text{liquid} \quad \Delta S_{\text{fusion}} = +5.258 \text{ cal/deg mole}$$
$$\text{Ice} \rightarrow \text{vapor} \quad \Delta S_{\text{sublimation}} = +44.674 \text{ cal/deg mole}$$
$$\text{Liquid} \rightarrow \text{vapor} \quad \Delta S_{\text{vaporization}} = +39.416 \text{ cal/deg mole}$$

For the reaction

$$2H_2 + O_2 \rightarrow 2H_2O_{\text{gas}} \text{ at } 298°K \text{ and } 1 \text{ atm total pressure}$$
$$\Delta S_{\text{combination}} = -21.21 \text{ cal/deg mole}$$

For the change

$$\text{Graphite} \rightarrow \text{diamond at } 298°K$$
$$\Delta S_{\text{transition}} = -0.7780 \text{ cal/deg mole}$$

For the reaction

$$\text{MgO (periclase)} + SiO_2 \text{ (quartz)} \rightarrow MgSiO_3 \text{ (enstatite) at } 298°K$$
$$\Delta S_{\text{reaction}} = -0.2 \text{ cal/deg mole}$$

We may draw a number of conclusions from these values. Entropy appears to measure the amount of order or chaos in a system, and thus entropy increases when dissociation, sublimation, melting, or mixing occurs, all being processes which lead to increased molecular chaos. In a solid-solid reaction, the change in entropy is small unless one set of solids is denser than another, and freedom is thus lost or gained as the volume containing the atoms decreases or increases. In most cases, consideration of which of two states is more ordered will allow a correct estimation of the sign of an entropy change.

We cannot go into the complexities of thermodynamics in this book, but it can be noted that all quantities needed to obtain G for a given substance in a given state are capable of direct measurement. If we differentiate equations defining ΔG of a reaction with respect to T at constant P and with respect to P at constant T, we obtain two important relations allowing the estimation of the change of ΔG with pressure and temperature.

These are:

$$\left(\frac{\partial \Delta G}{\partial P} \right)_T = \Delta V$$

$$\left(\frac{\partial \Delta G}{\partial T} \right)_P = -\Delta S$$

In words, these relations tell us that states of small volume (dense states) are favored by high pressures, while states of high entropy with greater atomic freedom are favored by high temperatures. These two equations are thus quantitative statements of part of Le Chatelier's principle. Our phase diagram of Fig. 1-1 is a representation of physical regions where a given state of a chemical system has a lower free energy, or chemical potential energy, than other possible states. Along phase boundaries, the adjacent states have equal free energies and are in equilibrium. It must be stressed that free-energy differences tell us nothing about the velocities of reactions, only whether or not they are possible.

Chapter 2

ATOMIC STRUCTURE

For most of the purposes of chemistry and the present discussion, the atom can be assumed to be built from three particles, the neutron, proton, and electron. The neutrons and protons are combined to form a nucleus which can be considered to be situated at a central point of the atom, carrying almost all the mass and having a positive charge equal to the number of protons—the atomic number. As the atom is neutral, a number of electrons, carrying little mass but negative charge equal to that of the proton number, circulate outside the nucleus. Chemical reactions do not involve the nucleus but touch only the outer fringes of the atom or the electrons, and thus in all chemical properties of matter we are concerned with the arrangement of these electrons, their ease of removal, spatial configuration, etc. We may assume that for each element (atom with a definite nuclear charge) we know the magnitude of the atomic number and neutron number and then proceed to see how the electrons are arranged.

At the outset, the fact that atoms exist at all is a puzzle. We are all familiar with the idea that unlike charges attract. Why then do the electrons not fall in on the nucleus? This may be seen more clearly if we consider the hydrogen atom, a bare proton with a single electron, mass m and charge e, moving in an assumed circular path of radius r at velocity v outside the nucleus (Fig. 2-1).

If this situation is stable, the force of attraction between the proton and electron e^2/r^2 must be balanced by the centrifugal force mv^2/r. Thus,

$$\frac{e^2}{r^2} = \frac{mv^2}{r} \tag{1}$$

10

The energy of the system is the sum of the kinetic energy $\frac{1}{2}mv^2$ and potential energy $-e^2/r$, that is,

$$E = \tfrac{1}{2}mv^2 - \frac{e^2}{r} \tag{2}$$

and as from (1) $m = e^2/rv^2$,

$$E = \frac{\tfrac{1}{2}e^2}{r} - \frac{e^2}{r} = \frac{-\tfrac{1}{2}e^2}{r} = -\tfrac{1}{2}mv^2 \tag{3}$$

This simple result poses the basic stability problem; our atom should become more stable as r decreases, and the electron should gradually fall onto the nucleus.

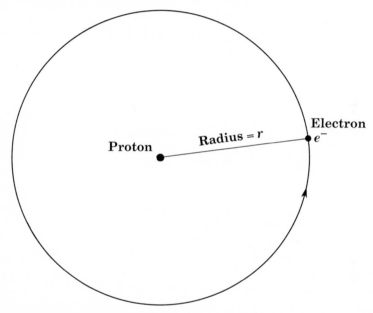

Fig. 2-1 Model of a hydrogen atom with an electron moving in a circular path around a central proton.

At the turn of this century, the German physicist Max Planck was worrying about blackbody radiation and the way radiation and matter interact. To describe the results, he was forced to introduce a new concept into physics, general but of absolute necessity in the physics of small particles. He postulated that when materials on the atomic scale change their energy, they do not do so continuously but only jump from one "quantum state" to the next. The idea of quantization of energy introduced by Planck was rapidly applied to other phenomena, such as the photoelectric effect by Einstein and, in 1913, to atoms by the Danish physicist Niels Bohr. It was known from the way atoms absorb and emit energy (from atomic spectra in the x-ray and ultraviolet regions of the spectrum) that these processes occur at definite wavelengths λ. There is, hence, nothing continuous about the energy of an atom. Bohr recognized that this again called for the application of Planck's quantum theory and sought a modification of the above treatment to accommodate quantum restrictions. He found that two postulates were sufficient to allow description of much of the hydrogen spectrum. First, he assumed that the angular momentum mvr of the electron could only take definite integral values; that is,

$$mvr = \frac{nh}{2\pi} \qquad n = 1, 2, 3, \ldots \tag{4}$$

h was a constant, Planck's constant, already found by Planck in his earlier study. From (4),

$$m = \frac{nh}{2\pi vr}$$

and from (1)

$$\frac{e^2}{r^2} = \frac{nhv}{2\pi r^2}$$

and

$$v = \frac{2\pi e^2}{nh}$$

Hence from (3) $\qquad E = -\tfrac{1}{2}mv^2 = \frac{-2\pi^2 me^4}{n^2 h^2} \cdots \tag{5}$

Bohr next postulated that when an electron jumps from one quantum state (stationary state) to another, the frequency v of the radiation absorbed or emitted was given by a relation due to Einstein:

$$E_1 - E_2 = hv$$

It had been known for some time that the wavelength λ of the major lines of the hydrogen spectrum could be described in terms of the relation

$$\frac{1}{\lambda} = R\left(\frac{1}{n_1{}^2} - \frac{1}{n_2{}^2}\right)$$

where $n_1 > n_2$, etc., were whole numbers and R was a constant. In the Bohr theory, $R = 2\pi^2 me^4/h^3 c$, where c is the velocity of light. It must surely have been considered a triumph when R calculated (109677.76) was compared with R observed (109678.18).

But there were still details in atomic spectra to be explained, and more quantum conditions were introduced, e.g., the restriction of types of possible elliptical orbits by Sommerfeld. Essentially this was a guessing game, but inherent in the game was the supposition that the behavior of individual electrons, such as their position, velocity, etc., could be measured or described. For example, we may use the above relations to show that the radius of the circular orbit of the hydrogen atom in its ground state of lowest energy is $h^2/4\pi^2me^2 = 0.53 \times 10^{-8}$ cm, the Bohr radius.

Early in this century these propositions began to crumble. De Broglie in 1924 suggested that particles and waves must be associated and suggested a general relation between the momentum mv of a particle and its wavelength λ:

$$mv = \frac{h}{\lambda}$$

Davisson, Germer, and Thompson experimentally confirmed this relation by showing that electrons showed diffraction effects similar to those of x-rays. Heisenberg proposed that it was meaningless and fundamentally unsound to talk simultaneously about such things as position and velocity for small particles, as the quantities are fundamentally incapable of determination. It is not difficult to see how this situation arises. Let us devise an experiment for measuring the position and velocity of a very small object. It is necessary that some particle interact with the system at two points and send us back a signal. Now if the particle is small, our messenger, say a photon, must disturb the particle quite drastically, and hence we cannot measure the true velocity. The problem is quite similar to that of determining the velocity and position of a railroad engine using a beam of artillery shells as the detector.

These and many other problems required and led to the development of a new approach to the problem of atomic structure, wave mechanics. Among those who contributed must be mentioned Dirac, Heisenberg, and Schrödinger. In the field of chemistry the most commonly used relation is the famous wave equation of Schrödinger. There is no need to discuss a formal derivation of this equation, for it must essentially be considered to stand alone as a fundamental relation, even though derivation by analogy with other wave systems is possible. As usually written, this equation is

$$\frac{\partial^2\psi}{\partial x^2} + \frac{\partial^2\psi}{\partial y^2} + \frac{\partial^2\psi}{\partial z^2} + \frac{8\pi^2m}{h^2}(E - V)\psi = 0$$

In essence, this is a differential equation which relates a quantity ψ, the wave function of the system, to the total energy E and potential energy V. Such an equation can be satisfied by an infinite number of values of ψ which lead to separate (not continuous) values of E for a given potential V. Of greatest interest are those solutions leading to the lowest possible values of E, the stable stationary states.

We cannot go too fully into the complete significance and restrictions involved in the solution of this equation and the meaning of

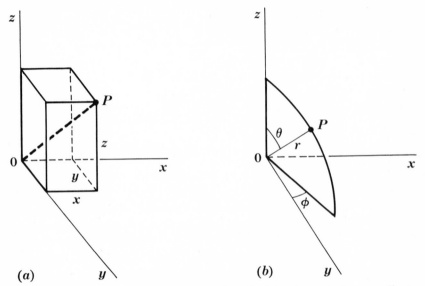

Fig. 2-2 The position of point P can be described by the cartesian coordinates xyz or spherical polar coordinates $r\theta\phi$.

ψ. But of great significance for our purposes is that the value of ψ^2 at any point in space is a measure of the probability of meeting the electron at that point; it is thus a measure of electron density or the number of electrons per unit volume. Before we proceed to look at the general results of the solutions to this equation, let us examine two examples to remove some of the apprehension regarding it and to see how the results in some directions are identical with those of the Bohr theory and in others totally different.

Many problems can be solved with much greater ease by the use of the equation in terms of spherical polar coordinates $r\theta\phi$ (see

Fig. 2-2), rather than *xyz* cartesian coordinates.   In these coordinates the equation takes the form

$$\frac{\partial^2 \psi}{\partial r^2} + \frac{2}{r}\frac{\partial \psi}{\partial r} + \frac{1}{r^2 \sin \theta}\frac{\partial}{\partial \theta}\left(\sin \theta \frac{\partial \psi}{\partial \theta}\right)$$

$$+ \frac{1}{r^2 \sin^2 \theta}\frac{\partial^2 \psi}{\partial \phi^2} + \frac{8\pi^2 m}{h^2}(E - V)\psi = 0$$

If we again consider the hydrogen atom, it is reasonable to suppose that there will be states where the electron density has spherical symmetry and the potential energy V will equal $-e^2/r$.   If there are spherical solutions, all derivatives depending on angular properties θ and ϕ will disappear, and the equation becomes

$$\frac{\partial^2 \psi}{\partial r^2} + \frac{2}{r}\frac{\partial \psi}{\partial r} + \frac{8\pi^2 m}{h^2}(E - V)\psi = 0$$

A spherical solution might be of the form

$$\psi = e^{-ra}$$

and

$$\frac{\partial \psi}{\partial r} = -ae^{-ra}$$

and

$$\frac{\partial^2 \psi}{\partial r^2} = a^2 e^{-ra}$$

Substitution of these first and second derivatives gives

$$a^2 - \frac{2a}{r} + \frac{8\pi^2 m}{h^2}\left(E + \frac{e^2}{r}\right) = 0$$

Now the relation must hold for all values of r, and hence

$$a^2 + \frac{8\pi^2 m E}{h^2} = 0 \tag{6}$$

and

$$-\frac{2a}{r} + \frac{8\pi^2 m e^2}{h^2 r} = 0 \tag{7}$$

From (7),

$$a = \frac{4\pi^2 m e^2}{h^2}$$

and hence from (6),

$$E = -\frac{2\pi^2 m e^4}{h^2}$$

exactly the Bohr solution for $n = 1$ [Eq. (5)].   This in fact is the solution for the hydrogen atom giving the lowest possible energy.

In Fig. 2-3 we have plotted the variation of ψ and ψ^2 the electron density as a function of r. It will be noticed that the greatest electron density is near the nucleus but that the concept of size has become vague; ψ^2 still has finite values at large distances from the nucleus. A useful quantity in discussing electron distribution around a nucleus is the function $4\pi r^2\psi^2(r)\,dr$. This function describes the number of electrons in a shell at radius r from the nucleus where ψ is the wave function at r and dr is the infinitesimal thickness of the shell. If we define a radial distribution function

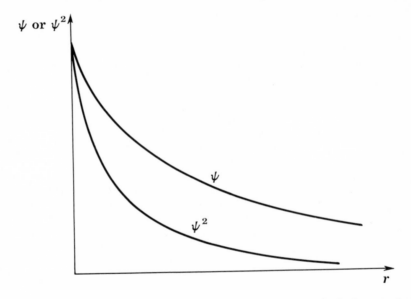

Fig. 2-3 Variation in the value of ψ and ψ^2 for the electron of a hydrogen atom as the distance from the nucleus increases.

D as $= 4\pi r^2\psi^2$, then when this function is plotted as a function of r, we see the variation in the electron density in shells as we pass out from the nucleus. For the hydrogen atom in its ground state this function is plotted in Fig. 2-4. It will be noticed that the function passes through a maximum where $dD/dr = 0$. We can easily locate this maximum.

$$\psi = e^{-ra}$$
$$D = 4\pi r^2 e^{-2ra}$$
$$\frac{dD}{dr} = 8\pi r e^{-2ra} - 8\pi r^2 a e^{-2ra} = 0$$
$$1 - ra = 0$$

and
$$r = \frac{1}{a} = \frac{h^2}{4\pi^2 me^2}$$

which is exactly the old Bohr radius. There is thus a striking correlation between the results of the two treatments, but the physical significance is vastly different.

There is an infinite family of spherical solutions of the form $\psi = e^{-cr}$ which give the same energies as the Bohr theory for circular orbits. These states we will call s states and an electron whose behavior is described statistically by such a function an s electron. The solutions of concern to us in atomic structure are those with the lowest energy values.

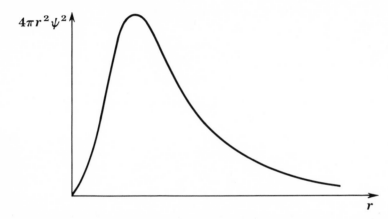

Fig. 2-4 Representation of the radial distribution function for the ground state of the hydrogen atom. The magnitude of $4\pi r^2 \psi^2$ at a given value of r represents the electron density in a thin spherical shell at distance r from the nucleus.

There are other solutions to this equation. For example, we could try a solution of the form

$$\psi = xe^{-cr}$$

where the x implies angular dependence in the x direction and lack of spherical symmetry. The solutions will only be of interest if the energies are comparable with some of the lower s states. For the hydrogen atom, the lowest energy of such states is almost exactly equal to that of the second spherical state, so these states are significant. Further, if there is a solution $\psi = xe^{-cr}$, there must be solutions $\psi = ye^{-cr}$, $\psi = ze^{-cr}$. Such a state, where there are three energetically equivalent, but spatially different, solutions, is described as being threefold or triply degenerate. States described by such a wave function are p states and the electrons in them p electrons. The electron distribution is concentrated axially.

We can proceed further and try the solution $\psi = xye^{-cr}$, and again there are solutions in the energy range of interest. The lowest state of this type is approximately equal in energy to the $3s$ state and $3p$ state. These d solutions are fivefold degenerate in that there are five solutions of equal energy. The only additional solutions of chemical interest show even more complex angular dependence and are the f states which are sevenfold degenerate, the lowest state being similar in energy to the $4s$, $4p$, $4d$ states. For the hydrogen atom we thus have a series of energy levels whose energies and terminology are summarized below.

$$- 5s \qquad\qquad \text{etc.}$$

$$- 4s \qquad\qquad \equiv 4p \qquad\qquad \equiv 4d \qquad\qquad \equiv 4f$$

$$E \uparrow \qquad - 3s \qquad\qquad \equiv 3p \qquad\qquad \equiv 3d$$

$$- 2s \qquad\qquad \equiv 2p$$

$$- 1s$$

A pictorial representation of the electron cloud for s, p, and d functions is shown in Fig. 2-5.

To build a model of the atom, we may proceed as follows. Electrons fed into an atom will occupy the lowest energy levels first, i.e., the $1s$, then $2s$, $2p$, etc. We need to consider two additional fundamental postulates: (1) Each unique quantum state (s, p_x, p_y, etc.) can accommodate only two electrons (but these must be spinning in opposite directions); (2) no two electrons in the same atom can be described by the same ψ unless their spins are opposed. These statements involve what is usually called the Pauli exclusion principle.

For complex atoms, the Schrödinger equation becomes extremely complex to solve. For example, the V term for an atom with 60 electrons must become impossibly complex. There are devices to avoid some of the difficulties but the actual energy levels of complex atoms have been derived from a combination of the wave equation and study of atomic spectra. Such an energy-level diagram is shown in Fig. 2-6 and electron configurations are given in Table 2-1.

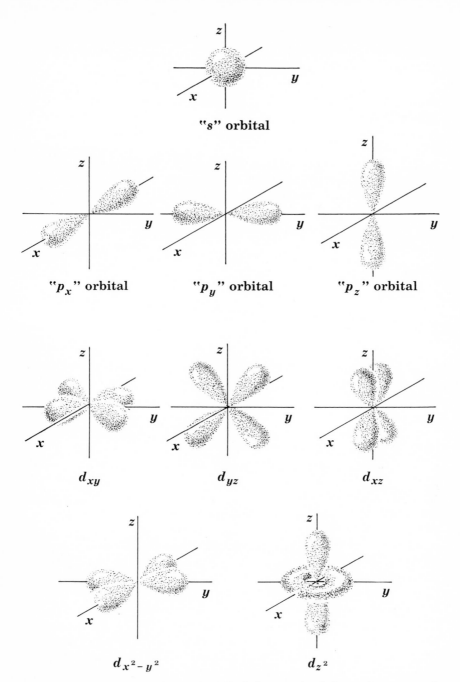

Fig. 2-5 Representation of the "shape" of electron clouds for various types of atomic orbitals.

19

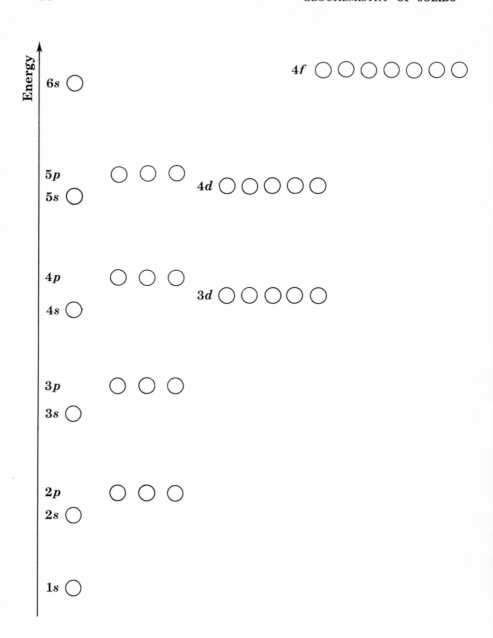

Fig. 2-6 Relative energies of atomic orbitals in atoms. Each state represented by ○ can accommodate two electrons with opposite spins, (↑↓).

Table 2-1   The Electronic Configurations of the Atoms of the Elements

Symbol	Atomic number, Z	Number and distribution of electrons						
		1s	2s 2p	3s 3p 3d	4s 4p 4d 4f	5s 5p 5d 5f	6s 6p 6d 6f	7s
H	1	1						
He	2	2						
Li	3	2	1					
Be	4	2	2					
B	5	2	2 1					
C	6	2	2 2					
N	7	2	2 3					
O	8	2	2 4					
F	9	2	2 5					
Ne	10	2	2 6					
Na	11	2	2 6	1				
Mg	12	2	2 6	2				
Al	13	2	2 6	2 1				
Si	14	2	2 6	2 2				
P	15	2	2 6	2 3				
S	16	2	2 6	2 4				
Cl	17	2	2 6	2 5				
A	18	2	2 6	2 6				
K	19	2	2 6	2 6	1			
Ca	20	2	2 6	2 6	2			
Sc	21	2	2 6	2 6 1	2			
Ti	22	2	2 6	2 6 2	2			
V	23	2	2 6	2 6 3	2			
Cr	24	2	2 6	2 6 5	1			
Mn	25	2	2 6	2 6 5	2			
Fe	26	2	2 6	2 6 6	2			
Co	27	2	2 6	2 6 7	2			
Ni	28	2	2 6	2 6 8	2			
Cu	29	2	2 6	2 6 10	1			
Zn	30	2	2 6	2 6 10	2			
Ga	31	2	2 6	2 6 10	2 1			
Ge	32	2	2 6	2 6 10	2 2			
As	33	2	2 6	2 6 10	2 3			
Se	34	2	2 6	2 6 10	2 4			
Br	35	2	2 6	2 6 10	2 5			
Kr	36	2	2 6	2 6 10	2 6			
Rb	37	2	2 6	2 6 10	2 6	1		
Sr	38	2	2 6	2 6 10	2 6	2		
Y	39	2	2 6	2 6 10	2 6 1	2		
Zr	40	2	2 6	2 6 10	2 6 2	2		
Nb	41	2	2 6	2 6 10	2 6 4	1		
Mo	42	2	2 6	2 6 10	2 6 5	1		
Tc	43	2	2 6	2 6 10	2 6 6	1		
Ru	44	2	2 6	2 6 10	2 6 7	1		
Rh	45	2	2 6	2 6 10	2 6 8	1		
Pd	46	2	2 6	2 6 10	2 6 10			
Ag	47	2	2 6	2 6 10	2 6 10	1		
Cd	48	2	2 6	2 6 10	2 6 10	2		
In	49	2	2 6	2 6 10	2 6 10	2 1		
Sn	50	2	2 6	2 6 10	2 6 10	2 2		
Sb	51	2	2 6	2 6 10	2 6 10	2 3		
Te	52	2	2 6	2 6 10	2 6 10	2 4		
I	53	2	2 6	2 6 10	2 6 10	2 5		
Xe	54	2	2 6	2 6 10	2 6 10	2 6		
Cs	55	2	2 6	2 6 10	2 6 10	2 6	1	
Ba	56	2	2 6	2 6 10	2 6 10	2 6	2	

Table 2-1 (*Continued*)

Symbol	Atomic number, Z	Number and distribution of electrons																		
		1s	2s	2p	3s	3p	3d	4s	4p	4d	4f	5s	5p	5d	5f	6s	6p	6d	6f	7s
La	57	2	2	6	2	6	10	2	6	10		2	6	1		2				
Ce	58	2	2	6	2	6	10	2	6	10	2	2	6			2				
Pr	59	2	2	6	2	6	10	2	6	10	3	2	6			2				
Nd	60	2	2	6	2	6	10	2	6	10	4	2	6			2				
Pm	61	2	2	6	2	6	10	2	6	10	5	2	6			2				
Sm	62	2	2	6	2	6	10	2	6	10	6	2	6			2				
Eu	63	2	2	6	2	6	10	2	6	10	7	2	6			2				
Gd	64	2	2	6	2	6	10	2	6	10	7	2	6	1		2				
Tb	65	2	2	6	2	6	10	2	6	10	9	2	6			2				
Dy	66	2	2	6	2	6	10	2	6	10	10	2	6			2				
Ho	67	2	2	6	2	6	10	2	6	10	11	2	6			2				
Er	68	2	2	6	2	6	10	2	6	10	12	2	6			2				
Tm	69	2	2	6	2	6	10	2	6	10	13	2	6			2				
Yb	70	2	2	6	2	6	10	2	6	10	14	2	6			2				
Lu	71	2	2	6	2	6	10	2	6	10	14	2	6	1		2				
Hf	72	2	2	6	2	6	10	2	6	10	14	2	6	2		2				
Ta	73	2	2	6	2	6	10	2	6	10	14	2	6	3		2				
W	74	2	2	6	2	6	10	2	6	10	14	2	6	4		2				
Re	75	2	2	6	2	6	10	2	6	10	14	2	6	5		2				
Os	76	2	2	6	2	6	10	2	6	10	14	2	6	6		2				
Ir	77	2	2	6	2	6	10	2	6	10	14	2	6	9						
Pt	78	2	2	6	2	6	10	2	6	10	14	2	6	9		1				
Au	79	2	2	6	2	6	10	2	6	10	14	2	6	10		1				
Hg	80	2	2	6	2	6	10	2	6	10	14	2	6	10		2				
Tl	81	2	2	6	2	6	10	2	6	10	14	2	6	10		2	1			
Pb	82	2	2	6	2	6	10	2	6	10	14	2	6	10		2	2			
Bi	83	2	2	6	2	6	10	2	6	10	14	2	6	10		2	3			
Po	84	2	2	6	2	6	10	2	6	10	14	2	6	10		2	4			
At	85	2	2	6	2	6	10	2	6	10	14	2	6	10		2	5			
Rn	86	2	2	6	2	6	10	2	6	10	14	2	6	10		2	6			
Fr	87	2	2	6	2	6	10	2	6	10	14	2	6	10		2	6			1
Ra	88	2	2	6	2	6	10	2	6	10	14	2	6	10		2	6			2
Ac	89	2	2	6	2	6	10	2	6	10	14	2	6	10		2	6	1		2
Th	90	2	2	6	2	6	10	2	6	10	14	2	6	10		2	6	2		2
Pa	91	2	2	6	2	6	10	2	6	10	14	2	6	10	2	2	6	1		2
U	92	2	2	6	2	6	10	2	6	10	14	2	6	10	3	2	6	1		2
Np	93	2	2	6	2	6	10	2	6	10	14	2	6	10	4	2	6	1		2
Pu	94	2	2	6	2	6	10	2	6	10	14	2	6	10	5	2	6	1		2
Am	95	2	2	6	2	6	10	2	6	10	14	2	6	10	6	2	6	1		2
Cm	96	2	2	6	2	6	10	2	6	10	14	2	6	10	7	2	6	1		2
Bk	97	2	2	6	2	6	10	2	6	10	14	2	6	10	8	2	6	1		2
Cf	98	2	2	6	2	6	10	2	6	10	14	2	6	10	9	2	6	1		2
Es	99	2	2	6	2	6	10	2	6	10	14	2	6	10	10	2	6	1		2
Fm	100	2	2	6	2	6	10	2	6	10	14	2	6	10	11	2	6	1		2
Md	101	2	2	6	2	6	10	2	6	10	14	2	6	10	12	2	6	1		2

This information is vital in all considerations of the properties of atoms and their compounds.

One further rule requires consideration. If a p, d, or f level is being filled, the filling follows the pattern indicated below with reference to the first transition-metal series. We shall designate each atomic orbital or atomic state such as $1s$ or $2p_x$ as $\bigcirc$, and this level can be singly occupied $(\uparrow)$ or doubly occupied $(\uparrow\downarrow)$, if the spins are opposed. The filling of the d level proceeds as indicated below:

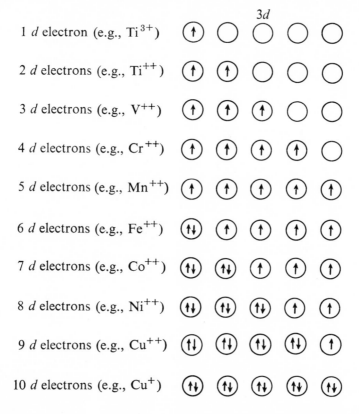

Thus in a degenerate level p, d, or f each state is singly occupied before it is doubly occupied—a reflection of the tendency of electrons to keep as far from each other as possible.

We are now aware of the way electrons are distributed in an atom and of the fact that various energy states are associated with different spatial distribution of the electron cloud. For further understanding of the chemistry of the elements it is important that we know more about the energy of addition and removal of electrons.

Let us take three examples of extreme types of behavior illustrated by fluorine, neon, and sodium. These form a group with the atomic structures:

	$1s$	$2s$	$2p$	$3s$
F	(↑↓)	(↑↓)	(↑↓) (↑↓) (↑)	( )
Ne	(↑↓)	(↑↓)	(↑↓) (↑↓) (↑↓)	( )
Na	(↑↓)	(↑↓)	(↑↓) (↑↓) (↑↓)	(↑)

The energies for the process of removal of one electron in the gas state, i.e., the process

$$X \rightarrow X^+ + e$$

are 17.42, 21.56, and 5.14 ev for F, Ne, and Na. (Note that 1 ev = 23,063 cal.) These quantities are known as the first ionization potentials. For the second step

$$X^+ \rightarrow X^{++} + e$$

the energies are 34.98, 41.07, and 47.3 ev. Clearly it is much easier to take away the first electron than the second, and further, sodium is much more readily ionized than fluorine or neon. As we know, fluorine is a reactive nonmetal, sodium a reactive metal, and neon inert, so a correlation begins to appear.

Next consider the process

$$X + e \rightarrow X^-$$

The energy of the process is a measure of the electron affinity E. The values here are −4.08 ev for fluorine, +1.2 ev for neon, and −0.2 ev for sodium. It will be noted that for fluorine and sodium a definite affinity exists but is large only for fluorine, whereas neon has a negative affinity for electrons. These two simple properties separate elements into three classes: (1) elements such as fluorine which readily acquire an added electron but do not easily lose electrons, (2) elements such as neon which do not easily lose or gain electrons, and (3) elements such as sodium which rather easily lose electrons and have little affinity for an added electron. The combination of these two properties I and E serves to distinguish metals from nonmetals and will tell much about the relative reactivity. The sum of these quantities may be used to estimate the electronegativity of an element which is a measure of an atom's tendency to gain or lose electrons. Fluorine is electronegative, sodium electropositive. In Table 6-2, electronegativity values for the common elements are listed. We shall consider further significance of these values in Chap. 6.

It may well be asked why such differences exist in ionization potentials and electron affinities. To offer an explanation of these values we must consider what is known as the "screening effect." Imagine that we know ψ and ψ^2 for each electron in a complex atom. With an atom such as sodium the difficulty of removing the outer electron is due to the attraction of the positive nucleus. But all the inner electrons repel the outer electron and effectively "screen" or neutralize part of this nuclear attraction. If we knew the inner electron distribution exactly, we could calculate the balance of the two effects. Thus the outer electron is in a resultant field due to a nuclear attraction we will call $Z_{\text{effective}} = Z_{\text{at. no.}} - S_{\text{electron screening}}$. A combination of theory and experiment has led to simple rules for calculating S. Some are given below.

1. Electrons are divided into groups of different screening capacity: $1s$; $2s$, $2p$; $3s$, $3p$; $3d$; $4s$, $4p$; $4f$; etc.
   2. S is the sum of the following terms:
      (*a*) Zero from all electrons outside the group considered.
      (*b*) 0.35 from each other electron in the group.
      (*c*) If the electron considered is s or p, 0.85 is removed for each electron in the next inner group and 1.0 from all electrons farther in.
      (*d*) If the electron considered is d or f, 1.0 is removed from all inner electrons.

Let us evaluate Z_{eff} for an outer electron in fluorine, neon, and sodium.
$$Z_{\text{eff}} \, F = 9 - 6 \times 0.35 - 2 \times 0.85 = 5.20$$
$$Z_{\text{eff}} \, Ne = 10 - 7 \times 0.35 - 2 \times 0.85 = 5.85$$
$$Z_{\text{eff}} \, Na = 11 - 8 \times 0.85 - 2 \times 1.0 = 2.2$$
This order is exactly that of the ionization potentials.

Now let us evaluate Z_{eff} for an added electron, i.e., the atom in the state X^-.
$$F^-, Z_{\text{eff}} = 9 - 7 \times 0.35 - 2 \times 0.85 = 4.85$$
$$Ne^-, Z_{\text{eff}} = 10 - 8 \times 0.85 - 2 \times 1.0 = 1.2$$
$$Na^-, Z_{\text{eff}} = 11 - 1 \times 0.35 - 8 \times 0.85 - 2 \times 1.0 = 1.85$$

Again the order is that observed for the electron affinities.

Thus with these rules and a knowledge of atomic structure, we can proceed a long way in predicting properties which influence chemical behavior.   Let us again stress that all these observations are found to follow from the use of one equation—the Schrödinger wave equation.

Chapter 3

FORCES BETWEEN ATOMS

The forces responsible for chemical combination of atoms are of two main types. There are those forces which arise between electrically charged species, repulsive if the charges are similar, attractive if dissimilar. These can be described as Coulomb forces and the relation

$$F = \frac{e_1 e_2}{r^2} \tag{1}$$

where r is the distance between the charges, and e_1 and e_2 the magnitude of the charges, will allow the calculation of the size of the force. The second type of force may be called an exchange force and these can be described in terms of the Schrödinger wave equation. The term *exchange* arises as follows. If two hydrogen atoms are separated at a large distance, we may identify an electron (1) with proton A and an electron (2) with proton B. Under these circumstances a wave function which will describe the whole system may be written as

$$\psi_{\text{system}} = \psi_A(1)\psi_B(2)$$

where $\psi_A(1)$ and $\psi_B(2)$ are the wave functions for the separate atoms. But as the atoms come closer together, and as the electron clouds begin to appreciably interpenetrate, this identification of an electron with a specific proton becomes vague, and the wave function for the system now becomes

$$\psi_{\text{system}} = \psi_A(1)\psi_B(2) + \psi_A(2)\psi_B(1) \tag{2}$$

Two terms must be used because neither electron can be considered to belong to a specific nucleus. The electrons may be considered to exchange protons, and this mathematical description has led to

27

the term *exchange interaction*. When the above ψ_{system} is used in the Schrödinger equation, the energy becomes lower than for the separate atoms; i.e., combination occurs.

It is common for people to feel that forces described by Eq. (1) are familiar and those described by (2) mysterious. A little reflection will reveal that we *understand* neither—all we have is two ways of describing different situations, and in all probability the origin of both forces is the same.

As well as these forces, we could include gravitational and magnetic forces, but both are so feeble in the discussion of chemical bonding that they may be neglected. For example, with a proton and an electron 10^{-8} cm apart, the Coulomb force is 2.3×10^{-3} dyne, whereas the gravitational force is 5.5×10^{-42} dyne.

Electrostatic forces in chemical compounds may be separated into the following groups:

Ion-ion interactions   e.g., Na^+Cl^- in NaCl

Ion-dipole interactions   e.g., Na^+—$O\begin{smallmatrix} H \\[2pt] \\ H \end{smallmatrix}$ in aqueous NaCl

Dipole-dipole interactions   e.g., $\begin{smallmatrix} H & & H \\ O & & O \\ H & & H \end{smallmatrix}$ in water

Induced dipole forces   e.g., solid helium
   (van der Waals forces)

We shall discuss dipoles at greater length in Chap. 6, but let us note here that in a molecule such as H_2O, while the entire molecule is electrically neutral, the charge resulting from nuclei and electrons is not symmetrically distributed, and the molecule has a polarity we can describe as

$$2\delta - \quad O \begin{smallmatrix} \overset{\delta+}{H} \\[4pt] \\ \underset{\delta+}{H} \end{smallmatrix}$$

where δ is some fraction of unit electric charge. The dipole is the product of charge separation δ and the distance between the charges.

Exchange forces are responsible for what are commonly called *covalent bonds*, e.g., the bond in the chlorine molecule Cl_2. Exchange forces lead to strong attraction only if one and preferably both atoms have at least one atomic orbital occupied by only one electron. Thus two chlorine atoms

$$1s \qquad 2s \qquad 2p \qquad 3s \qquad 3p$$

Cl Ⓤ Ⓤ ⓊⓊⓊ Ⓤ ⓊⓊ Ⓣ

attract each other strongly, two helium atoms

$$1s$$

He Ⓤ

$$1s$$

very feebly, but a helium atom and a positive helium ion (He⁺ Ⓣ) moderately strongly.

If a strong covalent bond is to form, i.e., if the electrons are to be effectively "exchanged," it is necessary that the charge clouds associated with the singly occupied levels coincide or "overlap" to a large extent. Thus in the hydrogen molecule, strong bonding occurs when the charge clouds are brought into close proximity as indicated in Fig. 3-1 below. As we shall see in more detail in Chap. 5, this necessity of overlapping imposes rigorous restrictions on the directional properties of covalent bands because some types of electron clouds are uniquely directed, a restriction which does not occur with electrostatic forces.

Whether the interaction is electrostatic or exchange in origin, all atomic molecular systems show similar distance-energy relations. We can illustrate this diagrammatically below (Fig. 3-2).

The repulsive and attractive energies add to produce a resultant with a distinct minimum which will correspond to the normal equilibrium internuclear distance. The attractive forces may be electrostatic or exchange. The repulsive forces arise from repulsion between the positively charged atomic nuclei and between the interpenetration of full electron shells. These forces tend to be short range or "hard" and increase quite rapidly as the atoms come close together. It is this rapid increase at short distances that leads to the incompressibility of solids and the total failure of the ideal gas equation

$$PV = RT$$

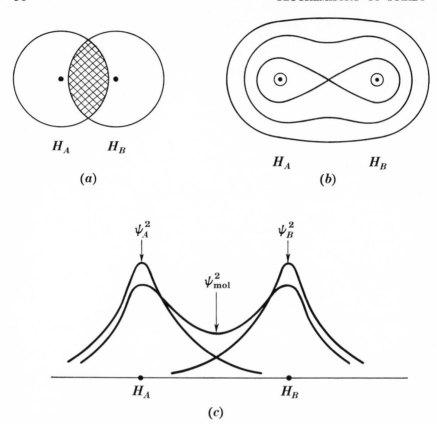

Fig. 3-1 Representation of overlap of atomic orbitals of hydrogen atoms in the formation of a hydrogen molecule: (*a*) overlap of essential boundaries of the clouds; (*b*) electron density contour map of a plane in the molecule, maximum density near the nuclei; (*c*) values of atomic densities ψ_A^2, ψ_B^2 and molecular densities ψ_{mol}^2 along the molecular axis. Note that formation of the molecule through exchange forces leads to a concentration of electron density between the nuclei.

at high pressures. Were it not for this, we would never be able to assign more or less constant sizes to atoms in compounds.

There are two useful ideas to keep in mind when discussing chemical bonds and electron distribution in molecules. First, as a result of exchange forces, electron spins tend to be correlated: electrons in singly occupied atomic states tend to occupy the same molecular state, i.e., spins become paired in molecules or compounds. Second, because of Coulomb forces, charges are correlated: electrons keep as far apart as possible, opposite charges as close together as possible.

Given any assemblage of atoms, it is useful to be able to predict what electron rearrangement may tend to occur and what binding forces may operate.   We can illustrate again with reference to fluorine, neon, and sodium.   When two fluorine atoms are brought together, the electrons in the two singly occupied p states will become paired, and strong exchange forces operate, resulting in the formation of a single covalent bond.   But each fluorine can only interact strongly with one other fluorine, and a diatomic molecule results.   When two neon atoms come together, no strong charge or spin correlation can act, and so the stable state is one of separate atoms.   When two sodium atoms approach in the gas phase, again a covalently bonded Na_2 molecule would be expected and is found.   If a sodium and fluorine atom come together, the two unpaired electrons will tend to couple, but this can happen in two ways.   A covalent bond may form, resulting in an Na—F molecule, or else the coupling occurs by electron transfer to form Na^+ and F^- because of the relative electron affinities of Na and F.   The ions will then interact

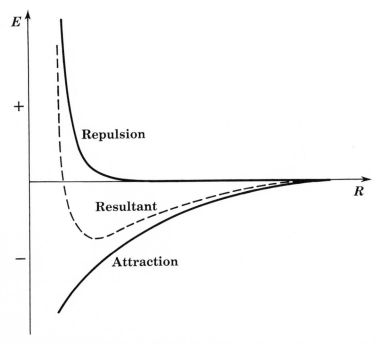

Fig. 3-2 Energy-distance relations in any system where atoms, ions, or molecules approach. Attraction, active at longer distances than the "hard" repulsion, produces a resultant with a minimum at the normal equilibrium distance.

through Coulomb forces to form the solid ionic compound NaF. But in the gas phase the more covalent molecule Na—F exists as well.

When a neon and fluorine atom come together, a molecule Ne—F could form via the strong exchange involving the states

Ne ($2p$) $\uparrow\downarrow$ F ($2p$) $\uparrow$ and   Ne ($2p$) $\uparrow$ F ($2p$) $\uparrow\downarrow$

but this cannot compete with the combination of two fluorine atoms, and hence the final state of the system can be represented as

$$Ne + F \rightarrow \tfrac{1}{2}F_2 + Ne$$

From the electronegativities of the atoms we can predict what type of bonding is likely to be dominant. If electronegativity differences are large, compounds will tend to be ionic (NaF); if zero covalent (F_2), and between these extremes, intermediate states will result. We may now proceed to discuss what properties are to be expected from ionic solids, covalent solids, and those in between.

Chapter 4

THE IONIC MODEL

When atoms whose electronegativities differ by a large amount form a solid compound, it may be assumed that extensive electron transfer will occur to form positive and negative ions. Thus, when sodium and chlorine form solid sodium chloride, a positive sodium ion and negative chloride ion will be involved because the atoms have such large differences in ionization potentials and electron affinities. We would now like to know how such ions will be arranged in a solid. The forces between the ions, Coulomb forces, can be described by a term e^2/r^2 and the energy by a term $-e^2/r$. One feature is immediately apparent. The requirement that the most stable structure have the lowest possible energy implies that the structure must be regular. Consider the very simple case of one cation and two anions. If the minimum distance of approach of cation to anion is r, we can readily calculate the energy as a function of the angle between the anions. The situation is illustrated in Fig. 4-1. The energy of this system equals twice the attraction between one anion and the cation minus the repulsion between anions:

$$E = -\frac{2e^2}{r} + \frac{e^2}{a}$$

$$= \text{const} + \frac{e^2}{a}$$

The energy is thus a minimum when a is a maximum, i.e., when $\theta = 180°$. This very simple example tells us much of what is to be found in more complicated examples. First, cations will want to be as close to anions as possible. Second, cations will want to be as far from cations, and anions from anions, as possible. Third,

33

certain definite angular arrangements will lead to the lowest energy. All these requirements must lead to extreme regularity.

Our simple example raises a very fundamental question: How close do the ions come together? We have already seen that around the nucleus of an atom we have a statistical electron cloud

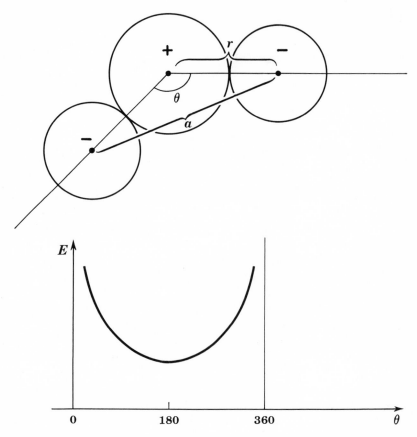

Fig. 4-1 Electrostatic energy of a three-ion system as a function of the angle θ, between the centers.

which becomes progressively diffuse out from the nucleus. Now electrons repel each other and tend to keep apart unless they occupy a single state and have opposed spins. Further, the positive nuclei repel each other. These charge correlation effects mean that as soon as the electron clouds of full states (e.g., all the electrons in Na^+ and Cl^-) begin to overlap appreciably, strong repulsion sets in and increases much more rapidly than the coulombic attraction.

When the rate of increase in attraction is less than the rate of increase of repulsion, penetration must cease.

Empirically, it is found that the situation can be described by an energy relation of the type

$$E = -\frac{e^2}{r} + \frac{e^2}{r^n}$$

The first term $-e^2/r$ is the attractive Coulomb energy. The second term $+e^2/r^n$ is the repulsion due to electron and nuclear repulsion.

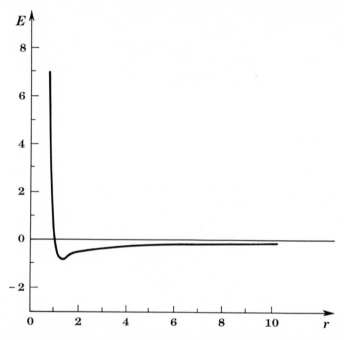

Fig. 4-2 The function $E = (-e^2/r) + (e^2/r^n)$ plotted for singly charged species as a function of r for $n = 10$. Note that the minimum value of E occurs when $dE/dr = 0$, that is, when $e^2/r^2 = 10e^2/r^{11}$ or when $r \simeq 1.3$. The extreme rise of repulsive forces at short distances should be noted.

We shall see later that the n in r^n has values between 8 and 12, so at large distances the repulsive term is small, but as r becomes small, e^2/r^n becomes extremely large. The situation is illustrated in Fig. 4-2. The resultant minimum of Fig. 4-2 will occur at the equilibrium distance. This curve, typical of all atomic-molecular systems, allows us to consider that atoms have a finite size, related to the electron distribution around the atom, a distribution which is sufficiently independent of the environment that to a first approximation we can talk of a more or less constant ionic size or ionic radius.

How is the size of an ion determined? Basically, study of the internal structure of crystals commenced when in 1912 von Laue suggested that if crystals had a regular structure, they should show diffraction effects with radiation of x-ray wavelengths, just as a ruled grating shows diffraction with light waves. Friedrich and Knipping showed that this was indeed the case, and W. L. Bragg and others developed the necessary treatment to determine the crystal structure and internuclear distance in crystals. Thus, from an x-ray diffraction pattern, the sum of the radii of two ions in a crystal may be estimated with a high degree of precision.

But we are left with a problem. We cannot obtain a crystal of all positive ions or negative ions by themselves. The two are always present together. How are we to divide the observed sum of the radii? As x-rays are scattered by electrons, it is in fact possible, by using refined analysis of x-ray scattering, to plot electron density contours in a given crystallographic plane, and from this the absolute sizes could be determined. The tables of ionic radii in use at present have not been determined by this method.

The method used in obtaining modern values is in outline as follows. Crystals of known simple structure with certain critical pairs of cations and anions are chosen. Such crystals are NaF, KCl, RbBr, CsI. It will be noticed that the ions in these crystals have the same electronic structure or are isoelectronic.

$$
\begin{array}{ll}
\text{Na} & 1s^2\, 2s^2\, 2p^6\, 3s^1 \\
\text{Na}^+ & 1s^2\, 2s^2\, 2p^6 \\
\text{F} & 1s^2\, 2s^2\, 2p^5 \\
\text{F}^- & 1s^2\, 2s^2\, 2p^6
\end{array}
$$

As the ions have identical electronic structures, we might expect them to be similar in size, but a little reflection indicates that this would be a poor guess. The position of the outer electrons is partly influenced by the attraction to the positive nucleus. As sodium has 11 protons and fluorine 9 protons in the nucleus, we would expect that Na^+ would be smaller than F^-. Thus our second guess would be that

$$
\frac{r_{Na^+}}{r_{F^-}} = \frac{9}{11}
$$

But this is still not the best division of the distance. It is the outer electrons that determine the chemical properties and size. The position of these electrons, as we have mentioned previously, is determined by the nuclear attraction and the inner-electron repulsion or screening. Thus the radius is going to be related to

$$Z_{\text{eff}} = Z - S$$

For the isoelectronic sodium and fluoride ions, $S = 4.52,$* and thus we have the two relations:

$$R = r_{\text{Na}^+} + r_{\text{F}^-} \quad \text{(measured)}$$

$$\frac{r_{\text{Na}^+}}{r_{\text{F}^-}} = \frac{9 - 4.52}{11 - 4.52}$$

This then gives a fluoride ion radius of 1.36 A (angstroms) and a sodium ion radius of 0.95 A.

If it is assumed that radii are strictly additive, in principle, only one pair need be treated as above to enable all radii to be determined. For example, given the radius of Na^+, a study of NaCl gives the radius of Cl^-; Na_2O, the radius of O^{--}; etc.

The assumptions in this treatment of additivity and essential spherical symmetry are not entirely reasonable. However, through their use a great deal of progress has been made in understanding ionic crystals, and this success indicates that the departures are not serious for many purposes. Radii for many common ions are listed in Table 4-1.

Having determined the size of an ion as indicated above and its charge from consideration of the electronic structure, we must now consider how these ions will fit together. Two general rules will help us. First, the structure we derive must contain ions in such a ratio that the crystal is electrically neutral. Second, maximum stability will be associated with regular arrangements which place as many cations around anions as possible (and vice versa) without putting ions with similar charges closer together than radii allow while bringing cation and anion as close as radii allow. In other words, we may treat the ions as spherical balls and pack them as closely as possible within the restriction of electric neutrality. We may illustrate with reference to a number of simple compounds.

If a model based on close packing of spheres is assumed, it is a simple proposition to calculate the ratio of the radii of the two types of spheres, A and B, which permits a certain number of B to fit around A, and vice versa. Normally only certain regular

---

* This value is not precisely that given by the general rules on p. 25 but is specially determined for these ions.

Table 4-1 Ionic Radii of Common Ions

Element	Oxidation state	Radius	Element	Oxidation state	Radius
Aluminum ...	3+	0.50	Manganese ...	2+	0.80
Antimony	5+	0.62		3+	0.66
	3+	0.90		4+	0.54
Arsenic	5+	0.47	Mercury	2+	1.10
	3+	0.69	Molybdenum .	4+	0.67
Barium	2+	1.35		6+	0.62
Beryllium	2+	0.31	Nickel	2+	0.72
Bismuth	5+	0.74	Nitrogen	3−	1.71
	3+	1.20	Oxygen	2−	1.40
Boron	3+	0.20	Potassium	1+	1.33
Bromine	1−	1.95	Rubidium	1+	1.48
Cadmium	2+	0.97	Selenium	2−	1.98
Calcium	2+	0.99	Silicon	4+	0.41
Cesium	1+	1.69	Silver	1+	1.26
Chlorine	1−	1.81	Sodium	1+	0.95
Chromium ...	2+	0.84	Strontium	2+	1.13
	3+	0.69	Sulfur	2−	1.84
	6+	0.52	Tellurium	2−	2.21
Cobalt	2+	0.74	Thallium	3+	0.95
Copper	1+	0.96		1+	1.40
Fluorine	1−	1.36	Thorium	4+	1.06
Gallium......	3+	0.62	Tin..........	4+	0.71
Gold	1+	1.37		2+	1.12
Hydrogen	1−	2.08	Titanium.....	4+	0.68
Indium	3+	0.81		3+	0.76
Iodine	1−	2.16		2+	0.90
Iron.........	2+	0.76	Tungsten.....	6+	0.67
	3+	0.64	Uranium.....	4+	0.97
Lead	2+	1.20	Vanadium....	4+	0.60
	4+	0.84	Zinc.........	2+	0.74
Lithium......	1+	0.60	Zirconium....	4+	0.80
Magnesium...	2+	0.65			

groupings are considered, and while these are by far the most common, other types are known.

Let us consider A to be very small compared with B; a limiting case would be a proton or hydrogen ion. Clearly only two anions can pack around such a small particle unless the bond be grossly lengthened (Fig. 4-3). This is obviously most unfavorable from inspection of the potential energy-distance relation of Fig. 3-2. Thus the lower limit of coordination numbers is 2. As soon as atom A reaches a certain critical size, three B atoms will surround it in a trigonal-planar group, the only regular threefold coordinated

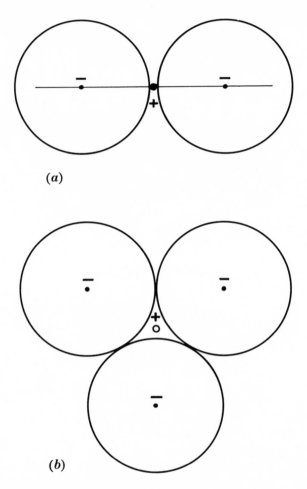

Fig. 4-3 Illustration of the limiting value of coordination where the cation is small. Generally, maximum energy is achieved when cation and anion touch, and situation B, where the cation rattles inside its coordinate group, is unfavorable.

structure.   The limiting value of the radius ratio for such packing is easily estimated (see Fig. 4-4).   The distance $b = 2R_B$, and distance $a = R_A + R_B$.

$$\sin \theta = \sin 60° = \frac{R_B}{R_A + R_B} = 0.866$$

$$0.134 R_B = 0.866 R_A$$

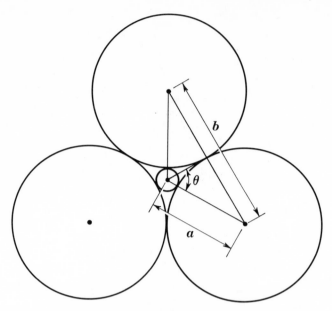

Fig. 4-4 Limiting geometry for trigonal planar coordination of atom A by three B atoms.   For explanation, see text.

and
$$\frac{R_{\mathrm{A}}}{R_{\mathrm{B}}} = 0.155$$

Critical radius ratio values for the most common type of coordination are summarized in Table 4-2.

It may well be asked why the numbers 5, 7, etc., are not considered. They do exist but are quite uncommon, for they do not lend themselves to the building of a large, regular three-dimensional framework.   When such numbers do occur, the bonds are generally quite covalent and the groups mixed with other groupings.

The sodium fluoride crystal can be considered to be made from sodium ions of radius 0.95 A and fluoride ions of radius 1.36 A. Quite clearly, it will be possible to pack many more sodium ions around fluorine than fluorine around sodium.   But the crystal must be neutral, and this necessitates that the coordination number of each ion be identical.   In any crystal of type A^+B^- this identity must hold, and the coordination number will be determined by the smaller ion, in this case, Na^+.   In the case of NaF, the radius ratio is 0.7 and the coordination number is 6.   A regular octahedral arrangement is expected and found.   This very common structure,

Table 4-2 Radius Ratios and Coordination Numbers

R_A/R_B	Coord. no.	Configuration		Example
0–0.155	2	Linear	F — H — F	$(HF_2)^-$
0.155–0.225	3	Trigonal planar		CO_3^{--}
0.225–0.414	4	Tetrahedral		SiO_2
0.414–0.732	4	Square planar		$Ni(CN)_4^{--}$
0.414–0.732	6	Octahedral		NaCl
0.732–1.0	8	Square bipyramid		CsCl

the simple cubic or sodium chloride arrangement, is common to many A—B compounds (Fig. 4-5 and Table 4-3).*

Table 4-3 Substances Crystallizing with the Sodium Chloride Structure

LiH	BaO	SnTe
LiF	BaS	SnAs
LiCl	ScN	PbS
LiBr	LaP	PbSe
LiI	UO	PbTe
NaH	UC	
NaF	TiO	
NaCl	TiC	
NaBr	MnO	
NaI	MnS	
KH	FeO	
KOH	CoO	
KF	NiO	
KCl	AgF	
KBr	AgCl	
KI	AgBr	
CsCl (above 445°C)	AgI	
MgO	CdO	
MgS	TlCl	
CaO	TlBr	
CaS	TlI	

If we consider CsCl, the radius of Cs^+ = 1.69 A, and the radius of Cl^- = 1.81. Again, it is the smaller ion Cs^+ which must determine the coordination numbers. The radius ratio in this case is 0.93 and a coordination number 8 is expected and found. The structure is known as body-centered cubic and again is common (see Table 4-4 and Fig. 4-6). With zinc sulfide (ZnS), the radii are Zn^{++} = 0.74, S^{--} = 1.84 and the radius ratio = 0.40. In this case the tetrahedral arrangement of Fig. 4-7 is found (see also Table 4-5).

* It should be stressed that these (and most) drawings of crystal structures represent only the geometry of arrangement of the centers of the atoms or the mean positions of the vibrating nuclei. The illustrations indicate a great deal of empty space, but a little reflection will indicate that electrons may be found anywhere in the volume of the crystal. But the electron density is concentrated near the nuclei and along directions to near neighbors. The extent or amplitude of vibrations of the atoms is quite large. With sodium chloride (Fig. 4-5) the sodium and chloride ions are vibrating about a mean position to the extent of 0.25 A at room temperature and at 600°C by 0.6 A.

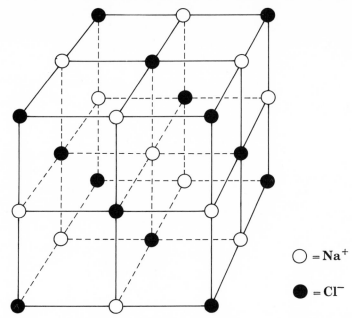

Fig. 4-5 The sodium chloride structure. Each ion is at the center of a regular octahedral group of oppositely charged neighbors.

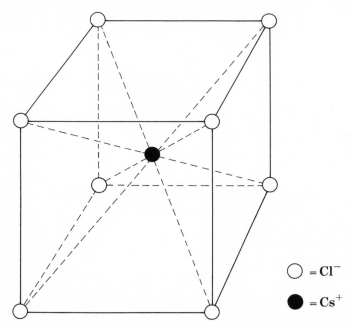

Fig. 4-6 The cesium chloride or body-centered cubic structure.

Table 4-4 Substances Crystallizing
with the Cesium Chloride Structure

CsCl (below 445°C)	SrMg*
CsBr	FeTi*
CsI	NiTi*
RbCl	CdAg*
RbBr	AgLi*
RbI	
NH₄Cl	

* Alloys, not ionic crystals.

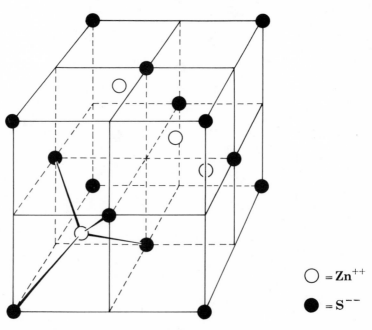

$\bigcirc = Zn^{++}$

$\bullet = S^{--}$

Fig. 4-7 The zinc blende (ZnS) structure. Each ion is in tetrahedral coordination.

Table 4-5 Substances Crystallizing with the Zinc Blende (ZnS, Sphalerite) Structure

BeS	ZnSe	HgTe
MnS	ZnTe	GaP
CuF	CdS	GaAs
CuCl	HgS* (meta cinnabar)	AgI
ZnO	HgSe	AlP
ZnS*		

* Note these sulfides occur in other modifications with a related structure.

With AB_2 compounds such as fluorite (CaF_2) or rutile (TiO_2), neutrality now necessitates that the coordination numbers be not identical. First, consider CaF_2 with R, Ca^{++} = 0.99 and R, F^- = 1.36. The radius ratio Ca^{++}/F^- = 0.72, so that a maximum number of six fluorines around Ca^{++} would be predicted, and this in turn means that only three calcium ions can surround fluorine. It will

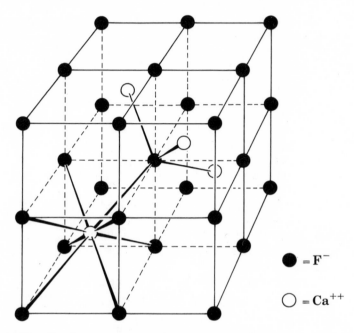

$\bullet = F^-$

$\bigcirc = Ca^{++}$

Fig. 4-8 The fluorite structure, each calcium surrounded by eight fluoride ions, and each fluorine by four calcium ions.

be noticed, however, that the ratio is quite close to the 6-8 crossing value. By a slight expansion of Ca—F bond, actually only by 0.005 A, the coordination number of 8 is possible, and if eight fluorines surround calcium, four calciums will surround fluorine, a situation which should give very much greater coulombic interaction. Thus the structure found is in 8:4 coordination (see Table 4-6 and Fig. 4-8).

Finally we may consider rutile (TiO_2). The radius of Ti^{4+} = 0.68 A, and O^{--} = 1.40 A. The radius ratio Ti^{4+}/O^{--} = 0.49, and six oxygens around titanium are predicted. The ratio is far from any crossover point and a 6:3 coordination number is anticipated with each Ti surrounded by an octahedral group of oxygens and each oxygen surrounded by three titaniums in a trigonal-planar configuration. This structure is illustrated in Fig. 4-9 (Table 4-7).

In the case of potassium chloride, the radius ratio of $K^+/Cl^- =$ 0.734, almost exactly on the 6-8 crossover value. It would be difficult to predict which structure would occur. Under ordinary conditions the 6-coordinated simple cubic modification is found.

Table 4-6 Substances Crystallizing with the Calcium Fluoride Structure

Li_2O^*	K_2O^*	CeO_2	Cu_2Se^*
Li_2S^*	Be_2C^*	ThO_2	Ag_2Te^*
Na_2O^*	CaF_2	UO_2	CdF_2
Na_2S^*	SrF_2	ZrO_2	HgF_2
K_2S^*	BaF_2	CuF_2	

* These substances where the cations and anions are in opposite positions to cations and anions in CaF_2 are sometimes said to have the antifluorite structure.

But the 8-coordinated form would lead to a packing more economical in space requirements, i.e., 8-coordinated KCl would be considerably more dense than the 6-coordinated modification. It would be anticipated, and has been experimentally observed, that when KCl is compressed to pressures of the order of 20,000 atm (about 290,000 psi), the 8-coordinated modification is formed. Quite generally, if ionic compounds are subjected to high pressures, structures with

Table 4-7 Substances Crystallizing with the Rutile Structure

MgF_2	WO_2	NiF_2
TiO_2	MnO_2	ZnF_2
VO_2	MnF_2	SeO_2
CrO_2	FeF_2	PbO_2
MoO_2	CoF_2	TeO_2

greater coordination numbers tend to form, a result of great interest in connection with deeply buried rocks and materials within the earth and other large bodies. Conversely, as the temperature is increased, smaller coordination numbers are to be anticipated. We shall return to consider such changes in Chap. 11.

With simple compounds like those considered above, the ionic model with consideration of radius ratios works well. But when we come to deal with more complex substances such as $CaCO_3$, where the anion CO_3^{--} is not spherically symmetrical but planar, some modification of the rules will be necessary. More complex types are considered in Chap. 9.

Almost all the properties and reactions of ionic crystals involve consideration at some stage of the energy to pull the crystal apart into its constituent ions at infinite separation. For example, if we wish to discuss the solubility of sodium chloride in water, it is

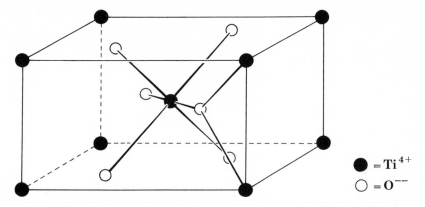

● = Ti^{4+}
○ = O^{--}

Fig. 4-9 The rutile structure, each Ti octahedrally coordinated and each oxygen surrounded by a trigonal-planar group of three Ti.

convenient to separate the process into parts: (1) the energy to take the lattice apart into free ions and (2) the energy of interaction of the water with the separated ions (solvation energy). If in the ionic model we are dealing with simple Coulomb forces, it should be possible to calculate this energy, frequently a dominant term in the free energy (see page 7).

It was found by the physicist M. Born that the potential energy of two ions in a crystal could be described by a relation of the form

$$V = -\frac{z_1 z_2 e^2}{r} + \frac{be^2}{r^n}$$

where z_1, z_2 = charges on cation and anion
r = distance between ions
b = const
n = const

As mentioned previously, the first term represents the normal coulombic attraction and the latter term the repulsion caused by interpenetration of the electron clouds with all electrons in doubly filled states plus the internuclear repulsion. Now to obtain the energy of the entire crystal, all such interactions must be added together over the entire crystal. This appears, at first sight, a formidable task. Consider a one-dimensional ionic crystal with interionic distance R:

$$\overbrace{\quad}^{2R}\ \overbrace{\quad}^{R}$$

$$\ominus\ \oplus\ \ominus\ \oplus\ \ominus\ \oplus\ \ominus\ \oplus\ \ominus\ \oplus\ \ominus\ \oplus\ \ominus\ \oplus\ \ominus$$

A

Now let us proceed to add all the interactions starting from atom A. For the attractive coulomb term e^2/r, we have

$$V = -\frac{2e^2}{R} + \frac{2e^2}{2R} - \frac{2e^2}{3R} + \frac{2e^2}{4R} - \frac{2e^2}{5R} \cdots$$

$$= -\frac{2e^2}{R}\left(1 - \frac{1}{2} + \frac{1}{3} - \frac{1}{4} + \frac{1}{5} \cdots\right)$$

It is apparent that the series

$$1 - \frac{1}{2} + \frac{1}{3} \cdots$$

is convergent, and thus by finding a suitable expression for the infinite sum, we can easily add all such interactions. The series limit in this case is ln 2 or $2.303 \log_{10} 2$. Thus we can write for the attractive interaction of atom A with all other ions in the linear array that

$$V = -\frac{\ln 2 \cdot e^2}{R}$$

It might be questioned why the infinite sum is used when the crystal is finite, but clearly all terms past, say $e^2/10^4 R$, are insignificant, and the sum will differ only extremely slightly from the infinite sum when we pass the first dozen terms, another way of saying that the coulombic interaction falls off quite rapidly with distance.

The summation of the repulsive terms proves to be even simpler. In this case all terms are positive, and

$$V = \frac{2e^2}{R^n} + \frac{2e^2}{(2R)^n} + \frac{2e^2}{(3R)^n} \cdots$$

From experimental studies of compressibility it is known that n has values in the range 8 to 12. If we consider atomic dimensions, R

will be of the order of 10^{-7} cm. If we put $n = 10$, the first term becomes

$$\frac{2e^2}{10^{-70}}$$

and the second term

$$\frac{2e^2}{(2 \times 10^{-7})^{10}} = \frac{2e^2}{1024 \times 10^{-70}}$$

Thus the second term is only one-thousandth of the first term, and clearly the third term

$$\frac{2e^2}{3^{10} \times 10^{-70}}$$

can be neglected. Thus to a good approximation, repulsion is only significant between nearest neighbors.

The Born equation for the entire crystal can be written as

$$V = -\frac{Ae^2 z_1 z_2}{R} + \frac{Be^2}{R^n} \tag{1}$$

The constant A, known as the Madelung constant, is the necessary summation constant, a function of the geometry or crystal structure. From the form of the potential energy curve (Fig. 3-2) it is clear that the most stable interionic distance occurs where the energy distance curve is at a minimum. Here we have

$$\frac{dV}{dR} = 0$$

If we differentiate (1) with respect to R (note if $y = x^n$, $dy/dx = nx^{n-1}$), we obtain

$$\frac{dV}{dR} = \frac{Ae^2 z_1 z_2}{R^2} - \frac{nBe^2}{R^{n+1}} = 0 \qquad \text{at equilibrium}$$

hence

$$\frac{Ae^2 z_1 z_2}{R^2} = \frac{nBe^2}{R^{n+1}}$$

and

$$B = \frac{Az_1 z_2 R^{n-1}}{n}$$

If this value of B is substituted into (1), we obtain

$$V = -\frac{Ae^2 z_1 z_2}{R} + \frac{Ae^2 z_1 z_2}{nR} = -\frac{Ae^2 z_1 z_2}{R}\left(1 - \frac{1}{n}\right)$$

To obtain the energy of the entire lattice array of 1 g mole of a crystal, this energy must be multiplied by N, Avogadro's number. The crystal energy or lattice energy U is defined as $-NV$, and hence

$$U = \frac{NAe^2z_1z_2}{R} \left(1 - \frac{1}{n}\right)$$

U represents the energy needed to take 1 g mole of the crystal apart into ions in the gas phase at infinite separation. To estimate the energy of the lattice, we must evaluate A, a constant dependent only on the geometry of the crystal and for which the values are known for all simple structures, and R, the interionic distance or sum of the radii. It is of interest to note that while n varies in the range 8 to 12, the value of $(1 - 1/n)$ is rather insensitive to n, for example,

$$\left(1 - \frac{1}{10}\right) = 0.9$$

$$\left(1 - \frac{1}{11}\right) = 0.9091$$

$$\left(1 - \frac{1}{9}\right) = 0.8889$$

Let us estimate the lattice energy of KCl with the above equation. We shall take $n = 9$; $R = R_{K^+} + R_{Cl^-} = 1.33\,A + 1.81\,A = 3.14\,A$, 3.14×10^{-8} cm; Avogadro's number $= 6.02 \times 10^{23}$; and e the electronic charge $= 4.80 \times 10^{-10}$ esu. The Madelung constant for this structure is 1.747. These figures will give the energy in ergs, and the conversion factor 2.389×10^{-11} is introduced to convert this to kilocalories (1000 cal). Hence

$$U = \frac{NAe^2z^2}{R}\left(1 - \frac{1}{n}\right) \times \text{conversion factor}$$

$$= \frac{\overset{N}{(6.02 \times 10^{23})}\overset{A}{(1.747)}\overset{e^2}{(4.8 \times 10^{-10})^2}\overset{z^2}{(1)^2}\overset{(1 - 1/n)}{(0.8889)}(2.389 \times 10^{-11})}{\underset{R}{3.14 \times 10^{-8}}}$$

$= 164$ kcal

It is impossible to measure the lattice energy of a crystal directly, but from related heats of reaction it can be estimated. This device makes use of Hess's law of heat summation, which says that in any process where the final and initial states are identical, the sum of all heats of reaction is zero. Thus in a process,

$$A \xrightarrow{\Delta H_1} B \xrightarrow{\Delta H_2} C \xrightarrow{\Delta H_3} D \xrightarrow{\Delta H_4} A, \quad \Delta H_1 + \Delta H_2 + \Delta H_3 + \Delta H_4 = 0$$

Consider the process outlined below:

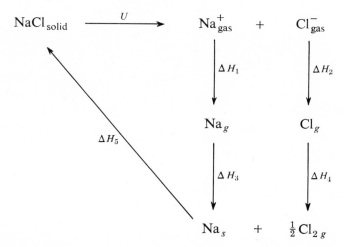

We have $U + \Delta H_1 + \Delta H_2 + \Delta H_3 + \Delta H_4 + \Delta H_5 = 0$, where

U = lattice energy
ΔH_1 = −ionization potential of sodium
ΔH_2 = −electron affinity of chlorine
ΔH_3 = heat of condensation of sodium
ΔH_4 = association energy of chlorine
ΔH_5 = heat of reaction of sodium and chlorine

Today, all the terms $\Delta H_1 \to \Delta H_5$ can be directly measured and U determined. The value found is 169 kcal, in reasonable agreement with the above estimate.

Returning to the equation

$$U = \frac{Ae^2 z_1 z_2}{R}\left(1 - \frac{1}{n}\right)$$

it is apparent that for crystals of the same structure, with ions of the same charge, the lattice energy varies as the interionic distance.

Thus $\qquad U_{\text{LiCl}} > U_{\text{NaCl}} > U_{\text{KCl}} \qquad$ etc.

For crystals containing ions of similar sizes we expect the energy to be related to the square of the charge, e.g., with LiCl, $R = 2.57$ A and with MgS, $R = 2.54$ A. We would expect that $U_{MgS} \simeq 4U_{LiCl}$. Experimental values are

$$U_{MgS} = 770 \text{ kcal} \qquad U_{LiCl} = 200 \text{ kcal}$$

Thus $\qquad U_{X^+Y^-} < U_{X^{++}Y^{--}} < U_{X^{3+}Y^{3-}}$ etc.

It is customary to relate many simple properties of ionic crystals to the lattice energy. We may mention a few cases. In the series NaCl, KCl, RbCl, all crystallize in the simple cubic structure, and as $R_{RbCl} > R_{KCl} > R_{NaCl}$, it follows that $U_{NaCl} > U_{KCl} > U_{RbCl}$. It is found that this is also the order of the melting points, 801°C for NaCl, 776°C for KCl, and 715°C for RbCl. On the other hand,

$$R_{NaCl} > R_{LiCl} \qquad \text{and} \qquad U_{LiCl} > U_{NaCl}$$

but the melting point of LiCl (613°C) is much lower than that of any other member of the series. While many examples do follow lattice-energy values, a great number do not. A little consideration indicates why the correlation should not necessarily be found.

At the melting point of a solid the free energy of liquid and solid forms must be equal. The temperature at which this equality is attained does not involve the energy of the crystal alone, but also the binding forces of the liquid. For example, at the melting point of KCl, the Coulomb forces between ions in the liquid are only 6000 cal less than in the solid, and the liquid is thus strongly bound. Factors which lead to a large Coulomb energy in the crystal will lead also to a large energy in the liquid, and it is not surprising that when the liquid is ignored, correlation with properties of only the solid are not good. Melting is a solid $\leftrightharpoons$ liquid equilibrium.

Hardness, a mechanical property of a crystal, shows rather better correlation, as might be expected, but again there are exceptions. Mechanical properties can be modified by various imperfections in a lattice (see Chap. 12). There are few simple properties of an ionic crystal which can be rigorously correlated with the lattice energy. This does not mean the quantity is unimportant. One final example will illustrate how the quantity appears in almost all chemical reactions involving ionic solids. Let us try to calculate the solubility of NaCl in water. In the solution, Na^+ and Cl^- ions are present but separated by layers of water molecules. To analyze the solution process, it is convenient to break the reaction into steps:

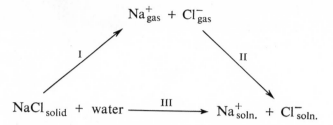

$$Na^+_{gas} + Cl^-_{gas}$$

$$NaCl_{solid} + water \xrightarrow{\quad III \quad} Na^+_{soln.} + Cl^-_{soln.}$$

Clearly the energy changes in I + II = III. With information on step I, we need only obtain information on step II to solve the problem. This process is quite complex, but an analysis of this type allows us to probe more intelligently into the factors which are significant in simple and fundamental processes.

We may finally attempt to answer a fundamental question in our study of solids. Why are the surfaces planar? We should perhaps note that the planar surfaces are rarely perfect, but contain steps; nevertheless, the major units are planar. It may seem a little surprising that this very obvious property of solids or crystalline materials arises from a rather trivial energy term—the surface energy of the substance. Consider the surface of a sodium chloride crystal as outlined below in a two-dimensional analogue.

$$
\begin{array}{cccccccc}
\overset{\text{A}}{\oplus} & - & + & \overset{\text{B}}{\ominus} & + & - & + & - \\
- & + & - & + & - & + & - & + \\
+ & - & + & - & + & - & + & - \\
- & + & - & + & - & + & - & + \\
+ & - & + & - & + & - & + & - \\
\end{array}
$$

It will be noticed that in the body of the crystal each ion has four close neighbors of opposite charge (six in 3 dimensions). In the surface we have two types of ions. Ions such as A in a corner position have lost two neighbors and two strong coulomb intersections. Ions such as B have lost one neighbor. In the three-dimensional model (Fig. 4-5), atoms at corners have lost three neighbors, on edges have lost two, and on faces one. In each case such an ion in the surface is less strongly bonded, and this lowers the stability of the solid. Clearly, if two shapes are possible, that with the smallest number of surface atoms or ions with unsatisfied

valence will be energetically preferred.   Compare, for example, the
square and rectangular two-dimensional crystals shown below, each
with 36 ions at a distance R and each with the same structure.

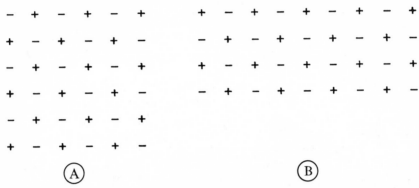

In structure A there are 20 surface ions, 4 corner and 16 edge.   In
structure B there are 22 surface ions, 4 corner and 18 edge.   Clearly
A will be more stable than B.   Thus we may conclude that quite
generally accentuated crystal habits, e.g., thin plates and thin needles,
should be uncommon.   One need only look at a mineralogy collection
to see that this is generally correct.   As discussed in Chap. 13,
rapidly growing crystals often have an extreme shape or habit, but
given time in a suitable medium, recrystallization leads to a more
nearly equidimensional habit.

   If the surface area is to be reduced to a minimum, it is clear that
a sphere is the most economical shape, but the regularity of the
atomic arrangement works in opposition to this advantage.   If a
crystal is to exhibit a perfectly spherical surface, a large number of
ions must occupy positions such as A and B shown below.

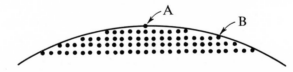

Such ions with only one or two valence forces satisfied are unstable
and would migrate to edge or corner positions transforming the
spherical surface into a series of planar units.   The regular arrangement thus leads to a planar surface.   If the regularity is destroyed,
and if a large number of the species gain freedom of translation as
in a liquid, the spherical drop has the lowest surface energy and is
preferred.

Chapter 5

THE COVALENT MODEL

Factors which influence the structures and physical properties of compounds where the bonding is covalent are quite different from those involved in ionic crystals. In the ionic model, the ions are considered to have finite size but are essentially structureless. When covalent models are considered, the atom can no longer be considered structureless or spherically symmetrical, and attention is focused on the nature and shape of the atomic orbitals containing unpaired electrons. In the bond, pairs of such orbitals on different atoms must overlap to provide the necessary exchange energy.

One general feature of covalent structures can be illustrated with reference to the elements silicon, phosphorus, sulfur, and chlorine. The atomic structure of chlorine

indicates only one electron suitable for exchange interaction with another chlorine; i.e., chlorine is monovalent. This electron in a p orbital has a shape (electron cloud) indicated previously in Fig. 2-5. When two chlorine atoms come together, attractive forces result only from interaction of electrons in these singly occupied orbitals. When these orbitals overlap appreciably, the wave function describing the electron distribution is a summation of parts of each atomic orbital. The valence electrons, one from each atom, must now be considered to move about both nuclei in what we can call a molecular orbital, and the wave function which describes their distribution is

$$\psi_{mol} = \psi_A + \psi_B$$

ψ_{mol}^2 describes the valence-electron density in the molecule and ψ_A^2

and $\psi_B{}^2$ the density in the separated atoms.   We can illustrate the situation as in Fig. 5-1.   Now, because only one unpaired electron per atom is available, a diatomic molecule is formed.   At low temperatures the thermal motion of the molecules is sufficiently reduced, and very weak polarization forces (see Chap. 7) lead to feeble binding of the molecules to form a solid at $-103°C$.   The

Fig. 5-1   Schematic representation of valence electron distribution in chlorine atoms and the Cl_2 molecule.

energy required to pull this solid apart into *molecules* is very small, and the structure that is found in the solid is based almost entirely on the closest packing of the dumbbell-shaped molecules.   These are packed together until the full electron shells begin to interpenetrate appreciably.   The forces binding such molecules are termed van der Waals forces, and we shall consider these in Chap. 7.

   The sulfur atom with the structure

1s	2s	2p	3s	3p
⊕	⊕	⊕ ⊕ ⊕	⊕	⊕ ⊕ ⊕

now has two orbitals singly occupied and hence can form two bonds. The possibility of one-dimensional polymerization now occurs.   On the basis of the electron distribution in the two singly occupied p orbitals, one might expect polymer shapes such as

$$
\begin{array}{cc}
S\!-\!S \\
| \quad | \\
S\!-\!S
\end{array}
\qquad
\begin{array}{cccc}
S\!-\!S & & S\!-\!S \\
| & & | \quad | \\
| & & S\!-\!S & S\!-\!S
\end{array}
$$

where all bonds are at 90°. The prediction on polymerization is correct, but the angular relations are more complex, as we shall see later in this chapter. Sulfur exists in several modifications; the orthorhombic form contains S_8 groups or molecules in the form of a ring, while plastic sulfur contains chains of atoms. A more stable solid than solid chlorine results, and the melting point now reaches 113°C. This increase in melting point compared to chlorine does not reflect any increase in the energy of the S—S bonds, for these still exist in the vapor and liquid state, again in complex arrangements. The chains, rings, etc., are still bound by van der Waals forces, and the increased thermal stability of the solid is more a reflection of the increased effective molecular weight or what we might term a *thermal inertia* of the molecules.

Phosphorus has the atomic structure

The possibility now exists of two-dimensional polymerization, for each phosphorus can be linked to three other phosphorus atoms. An ideal structure might be as shown below:

Such a structure is observed, though buckled, in black phosphorus which has a layer lattice. As with sulfur, other forms exist also, in particular a form with P_4 molecules arranged to form a regular triangular-based pyramid. The higher degree of polymerization is again reflected in the melting point of nearly 600°C.

Silicon has the atomic structure

At first sight a dicovalent silicon might be expected, but a new possibility arises with this element. As the $3s$ and $3p$ levels are quite similar in energy, combination may also occur with the atom in the "excited state":

In this state four bonds can be formed, and if the energy of formation of these four bonds is sufficient to supply the excitation energy, i.e.,

Energy of four bonds $>$ energy of two bonds $+$ excitation energy

Fig. 5-2 Tetrahedral arrangement of bonds in solid silicon. This is the diamond structure.

this four-covalent state will be favored. As is well known, both carbon and silicon are characteristically 4-covalent. The silicon in the excited state can now be linked to four other silicons, and a three-dimensional polymer can be formed. This must lead to a much stronger giant molecule crystal, as with all ionic crystals. The melting process must now involve some bond breaking, and the melting temperature for silicon reaches 1420°C. The structure is shown in Fig. 5-2, where each silicon is surrounded by four neighbors in a tetrahedral configuration.

We thus conclude that the physical properties of covalent substances reflect the number of bonds much more than the strength of the bonds. The bond energies in the cases discussed, i.e., the energy to break Cl_2 into Cl atoms and to break one S—S bond, etc., are

Cl—Cl	58,000 cal
S—S	50,900 cal
P—P	51,300 cal
Si—Si	42,200 cal

almost an inverse correlation with the physical properties. Again, the number of covalently bonded neighbors is unrelated to the covalent radii, which are quite similar.

Atom	Covalent radius, A	Near neighbors
Cl	0.99	1
S	1.04	2
P	1.10	3
Si	1.17	4

On a close-packing model all would take at least eight near neighbors.

There are no difficulties in the evaluation of covalent radii, which are again remarkably consistent for the same reasons as ionic radii. If the carbon-carbon distance in diamond is measured, this distance halved gives the covalent radius of carbon. The consistency is illustrated in Table 5-1, and values of covalent radii are given in Table 5-2. Covalent radii are normally listed for a given spatial configuration. The reasons for this will become apparent below.

We should note that covalent radii differ in a simple manner from ionic radii:

$$Cl_{ionic}^-, R = 1.81 \text{ A} \qquad Cl_{covalent}, R = 0.99 \text{ A}$$
$$Si_{ionic}^{4+}, R = 0.41 \text{ A} \qquad Si_{covalent}, R = 1.17 \text{ A}$$

The covalent chlorine is in a state where one outer electron has been drawn away toward its partner; in the ionic chlorine, an electron has been added. With the silicon ion, the atom has been entirely stripped of its $3s$ $3p$ electrons, and the nuclear pull on the remainder is so large that a very small ion exists. We shall discuss the significance of these facts again in Chap. 8.

Up to this stage the question of the angular distribution of covalent bonds has been deliberately neglected. As stressed previously, to

Table 5-1   Carbon-Carbon Distance in Some Carbon Compounds

Diamond		1.542
Ethane	H_3C—CH_3	1.533
Propane	H_3C—CH_2—CH_3	1.54
n-Butane	CH_3—CH_2—CH_2—CH_3	1.534
Cyclohexane	H_2C⟨... ⟩CH_2 (ring)	1.53

```
                  H2 H2
                  C—C
                 /    \
      H2C            CH2
                 \    /
                  C—C
                  H2 H2
```

Examples of additivity of radii

$\frac{1}{2}$ × C—C distance in diamond	0.77
$\frac{1}{2}$ × Cl—Cl distance in chlorine molecule	0.99
Observed distance in CCl_4	1.76
Sum of $\frac{1}{2}$ distances	1.76
$\frac{1}{2}$ × Si—Si distance in silicon	1.17
$\frac{1}{2}$ × Cl—Cl distance in chlorine	0.99
Observed distance in $SiCl_4$	2.01
Sum of $\frac{1}{2}$ distances	2.16

obtain strong exchange interaction, or strong covalent bonds, the electron clouds must overlap or the wave functions coincide in space. It seems clear with some simple molecules what configurations this should lead to. Consider the molecules hydrochloric acid, hydrogen sulfide, and phosphine. In each case the hydrogen uses a $1s$ orbital and the other atom 1, 2, or 3 p orbitals. The resulting situations are indicated in Fig. 5-3. In these cases the measured bond angles are very close to those predicted in this fashion (H_2S, 92.2°; H_3P, 93.3°). We shall consider the cause of the small but very significant deviations below.

A difficulty arises when we come to discuss quadrivalent states of carbon and silicon. As mentioned above, when silicon combines with other elements, the ground state $1s^2\,2s^2\,2p^6\,3s^2\,3p^2$ changes to the excited or valence state $1s^2\,2s^2\,2p^6\,3s^1\,3p^3$; the same is true for carbon $1s^2\,2s^2\,2p^2$ which changes to $1s^2\,2s^1\,2p^3$. What will happen when such an "excited" carbon atom interacts with four hydrogens to form CH_4 (methane)? The four singly occupied states are not

Table 5-2   Covalent Radii

Element	Configuration	Radius, A
Aluminum	Tet	1.26
Antimony	Tet	1.36
Arsenic	Tet	1.18
Beryllium	Tet	1.06
Boron	Tet	0.88
Bromine	Tet	1.11
Cadmium	Tet	1.48
Carbon	Tet	0.77
Chlorine	Tet	0.99
Cobalt(III)	Oct	1.22
Cobalt(II)	Oct	1.32
Copper	Tet	1.35
Fluorine	Tet	0.64
Gallium.	Tet	1.26
Gold(IV).	Oct	1.40
Hydrogen	. . .	0.30
Indium	Tet	1.44
Iodine	Tet	1.28
Iron(II)	Oct	1.23
Mercury	Tet	1.48
Nickel(II)	Oct	1.39
Nitrogen	Tet	0.70
Oxygen	Tet	0.66
Phosphorus.	Tet	1.10
Selenium	Tet	1.14
Silicon.	Tet	1.17
Silver.	Tet	1.52
Sulfur	Tet	1.04
Tellurium	Tet	1.32
Tin.	Tet	1.40
Titanium(IV)	Oct	1.36
Zinc.	Tet	1.31

Note: In the case of an element such as carbon, there is no problem in deciding on a compound to use for determination of the covalent radius. With an element such as aluminum, where the pure element is a metal, not a simple covalent compound, some selection is required. The compound chosen should be one where the difference in electronegativity of aluminum and its neighbors is small and the compound a poor conductor of electricity. Suitable compounds might be the nitride AlN, or an alkyl such as triethyl aluminum ($Al(C_2H_5)_3$).

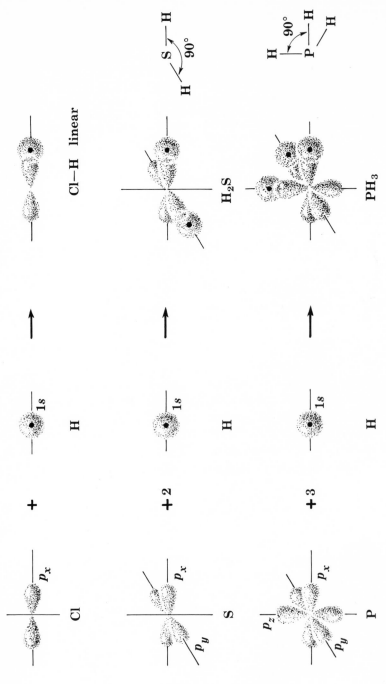

Fig. 3-5 A schematic diagram indicating the angular bond distribution in HCl, H_2S, and PH_3 as might be expected from combination of s and p orbitals.

similar. There are a vast number of chemical reasons which insist that the four bonds are equivalent, and wave mechanics again tells us that the lowest possible energy of the system will be achieved if these four orbitals, $3p$ and $1s$, rearrange or "hybridize" in the molecule. It is found that the electron distribution around the carbon nucleus can be described by a function:

$$\psi_{\text{hybrid}} = \psi_{2s} + \sqrt{3}\ \psi_{2p}$$

The separate s and p functions are mixed. The directional properties of the four equivalent (degenerate) new orbitals are tetrahedral, and the methane molecule is exactly tetrahedral as in Fig. 5-4. In

Table 5-3 Hybrid Orbital Configurations

Molecule	Orbitals on A	Hybrid	Configuration	Example
AB_2	s and p	sp	Linear	CO_2
AB_3	$s + 2p$	sp^2	Trigonal-planar	C_2H_4
AB_4	$s + 3p$	sp^3	Tetrahedral	CH_4
AB_4	$d + s + 2p$	dsp^2	Square-planar	$Ni(CN)_4{}^{2-}$
AB_6	$2d + s + 3p$	d^2sp^3	Octahedral	SF_6

diamond, or solid silicon, the atoms are in this state, and because of necessity of maximum overlapping, a tetrahedral structure results. The diamond structure is illustrated in Fig. 5-2.

There are many common types of hybrid mixtures which give rise to characteristic molecular configurations. Some are listed in Table 5-3.

We can now begin to consider the structures of most simple molecules. For example, in sulfur the long chains of S—S—S—S atoms form from linear sp hybrids. The solid sheet structure of graphite results from sp^2 trigonal-planar hybrids.

There is still one additional form of covalent exchange interaction we must consider, and this is well illustrated by the other form of elemental carbon, graphite. Graphite has a planar structure as indicated in Fig. 5-5. The atoms are arranged in rings with each carbon linked to three other carbons, the bond angles being exactly 120°, angles to be expected from sp^2 hybrids. Now let us examine a single carbon atom (Fig. 5-6). The bonds which are responsible for the 120° bond angles are formed from carbon sp^2 trigonal-planar hybrids. But a p orbital, not used in the hybridization and containing a single electron, is left with its lobes above and below

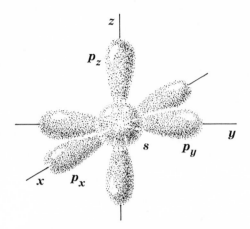

Separate s, p_x, p_y, p_z orbitals

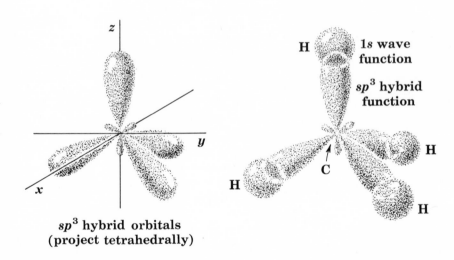

sp^3 hybrid orbitals
(project tetrahedrally)

1s wave
function

sp^3 hybrid
function

Fig. 5-4 Illustration of the shape of an sp^3 tetrahedral hybrid orbital and the configuration of the methane molecule.

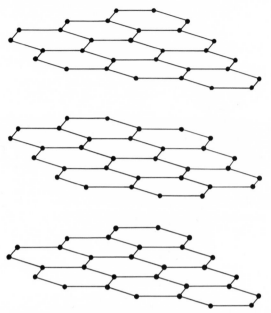

Fig. 5-5 Structure of graphite where carbon atoms are in an sp^2 hybrid configuration. Only weak forces operate between layers.

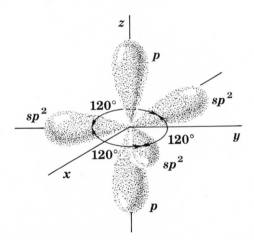

Fig. 5-6 The carbon atom in an sp^2 state. The three lobes of the sp^2 hybrid lie in a plane, the p lobe normal to the plane.

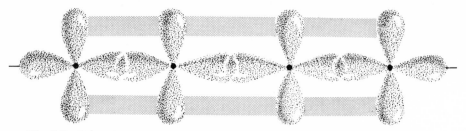

Fig. 5-7 Edge view of carbon atoms in graphite.   Along the line of atoms the *sp²* hybrids overlap.   The *p* orbitals project above and below the plane of the carbon layer.

the plane of the *sp²* hybrid orbitals.   Thus if we look edge-on at a carbon layer in graphite, we shall see a view as in Fig. 5-7.

These unpaired *p* electrons can achieve a more stable configuration if the spins become paired, and this can be achieved by a sideways overlap as in Fig. 5-8*a*.   The resulting electron distribution in the molecular orbital formed, $\psi_{mol} = \psi_{px} + \psi_{px}$, is shown in Fig. 5-8*b*.

The bond formed is called a π bond and the electrons have become "delocalized."   In the graphite structure, the *p* electrons on each carbon atom interact equally with the three neighboring *p* electrons, and the net result is that the delocalized π orbitals formed are smeared out through planar layers of the entire crystal.   The classical example of such bonding is shown by benzene, and the π electron distribution is shown schematically in Fig. 5-9.   The normal bonds formed by end-on overlap of orbitals are termed σ bonds to distinguish them from π bonds formed by edge overlap.   All stages in this development are seen in the compounds ethane (C_2H_6), ethylene (C_2H_4), and acetylene (C_2H_2) (Fig. 5-10).   We could

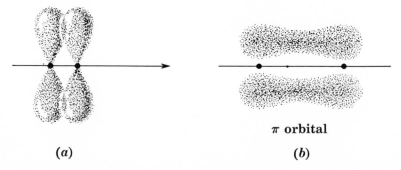

π **orbital**

(*a*)   (*b*)

Fig. 5-8 (*a*) Side overlap of two p_x orbitals to produce a π bond with delocalized electrons; (*b*) the resulting electron distribution caused by π overlap. The electrons in the π orbital have become "delocalized."

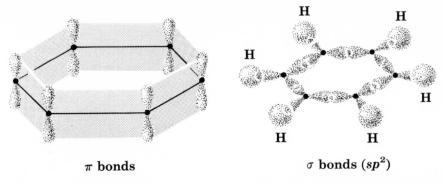

π bonds   σ bonds (sp^2)

Fig. 5-9 Representation of σ and π orbitals in the planar benzene molecule (C_6H_6). The σ bonds are formed from carbon sp^2 hybrids. The π electrons are smeared out above and below the plane of the carbon ring.

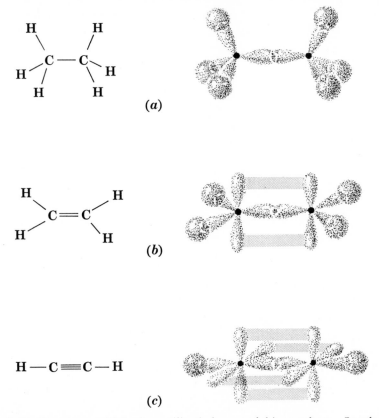

(a)

(b)

(c)

Fig. 5-10 Structures of (a) ethane, (b) ethylene, and (c) acetylene. In ethane, carbon is in the normal sp^3 hybrid state forming four σ bonds. In ethylene, the carbon is in an sp^2 hybrid state forming three σ bonds and one π bond. In acetylene, carbon is in an sp state forming two σ bonds and two π bonds.

67

represent these in terms of classical structures containing single, double, and .triple bonds (Fig. 5-10). Similar structures can be drawn for graphite, but here it must be stressed that all bonds are equivalent, and the π electron cloud is continuous throughout the crystal.

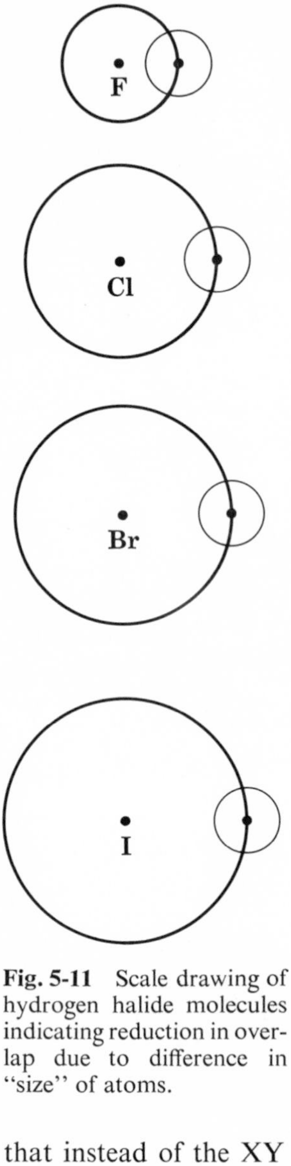

As might be suggested by the term *delocalized* applied to π electrons, these electrons are much more mobile and hence influenced by electric disturbances. In fact, graphite is quite a good conductor of electricity, while diamond with only σ bonds is a very good insulator. In describing the structure of a covalent compound, we thus need to know:

1. The number of unpaired electrons and the accessibility of excited states. These factors will determine the valence or coordination numbers.

2. The type of hybrids which the bonding orbitals may form.

These will determine the geometry of the arrangement in the molecule or crystal. Only occasionally will size appear as a critical factor, and when it does, it may often manifest itself not in a change in coordination but in the complete instability of a compound. An excellent example of this is provided by the compounds HF, HCl, HBr, and HI. The bond energies of these compounds, i.e., the energy of breaking the bonds and forming atoms, are 147.5 kcal HF, 102.7 kcal HCl, 87.3 kcal HBr, 71.4 kcal HI. Let us draw scale models of those atoms and examine the overlapping (Fig. 5-11). Quite clearly the overlap of hydrogen with fluorine can be much greater than of hydrogen with iodine, and this effect can become so extreme

Fig. 5-11 Scale drawing of hydrogen halide molecules indicating reduction in overlap due to difference in "size" of atoms.

that instead of the XY molecule forming, the elements X_2 and Y_2 represent a more stable state. To form a strong covalent bond, the charge clouds of the atoms in a bond must be of similar size.

An atom like carbon may occur in many hybrid states, and in each state the radius and electronegativity of the atom changes. Thus, in assigning covalent radii, we must specify the state of hybridization in the compound (Table 5-4). For carbon the radii are 0.772 sp^3, 0.665 sp^2, 0.602 sp.

Table 5-4 Covalent Radii, A, of Atoms Where Multiple Bonds Are Present

Carbon:		Phosphorus:	
Triple	0.602	Single	1.10
Double	0.665	Double	1.00
Single	0.772	Triple	0.93
Silicon:		Chlorine:	
Single	1.17	Single	0.99
Double	1.07	Double	0.89
Triple	1.00		

It should again be emphasized that there is no inherent relation between the strength of covalent bonds and the physical properties of the covalent solid. The strength of covalent bonds is quite comparable with ionic bonds, but very often in the covalent solid the strong bonds are only active within molecules and not between molecules, a situation which does not arise in ionic compounds. When a giant molecule is built, as in diamond, silicon, silicon carbide, etc., the solid is quite as strong as any ionic crystal.

The Coordinate Bond—Complex Ions

There is another form of strong exchange interaction involved in what are commonly called *coordinate bonds*. In most ways this bond can be considered as a covalent linkage with one distinction: When the coordinate bond is ruptured, both electrons in the bond go away with one species. When the hydrochloric acid molecule is ruptured, the fission products are H_2 and Cl_2 or H and Cl, and the bonding electrons are equally shared. When a hydrate such as $NiSO_4 \cdot 6H_2O$ is dehydrated, the covalent bonds between Ni^{++} and water are broken, but both bonding electrons remain with water molecules. In $NiSO_4 \cdot 6H_2O$, the structure shows us that each nickel ion is surrounded by six water molecules, and the Ni—O distance is about that to be anticipated from a covalent linkage. The current picture of this linkage is approximately as follows. On each oxygen surrounding the nickel ion there are two pairs of electrons not used in bonds to hydrogen. When the water molecule approaches the ion, the electron cloud representing one of these lone pairs interacts with an "empty" orbital in the metal and forms a covalent bond—abnormal only in the sense that both electrons have come from oxygen. The nickel ion uses a d^2sp^3 hybrid to form six such bonds octahedrally directed. A clear-cut case of such interaction is shown by the complex ion $Co(NH_3)_6^{3+}$ where a cobaltic ion is linked to six ammonia molecules, each with one pair of electrons not used in bonds to hydrogen. The structure of the outer electrons of the cobalt atom is

and when the ion is formed, this becomes

$$Co^{3+} \quad (\uparrow\downarrow) \; (\uparrow) \; (\uparrow) \; (\uparrow) \; (\uparrow) \qquad \bigcirc \qquad \bigcirc \; \bigcirc \; \bigcirc$$

Magnetic data show that in the complex no unpaired electrons remain. Thus a new state

$$(\uparrow\downarrow) \; (\uparrow\downarrow) \; (\uparrow\downarrow) \left[(:) \; (:) \quad \underset{d^2sp^3}{(:)} \quad (:) \; (:) \; (:) \right]$$

is formed, leaving a set of six empty orbitals to accommodate six electron pairs from the nitrogen atoms of the ammonia molecules in an octahedral configuration.

These bonds are formed with great frequency between cations with large ionization potentials and anions or molecules with electron pairs not used in bonding. Again, the coordination numbers are not controlled by size but more by choice of hybrid orbitals. Some examples are listed on page 71.

The strength of the coordinate bond appears to be influenced by two major factors: (1) it increases with increased ionization potential of the metal; (2) it increases with decreased ionization potential of the donor anion or molecule. These rules are too simple to cover all the aspects of these bonds but apply in a large number of cases. Thus frequently iodide complexes are more stable than chloride or bromide complexes, ammonia complexes more stable than water complexes, etc.   Both the above factors seem reasonable when it is considered that in the formation of the bond, electrons are drawn away from a donor and placed in orbitals of the acceptor cation.

Cation size, A		Complex	Metal hybrids
Ag^+	1.26	$Ag(NH_3)_2^+$	sp linear
Ag^{++}	<1.26	$Ag(NH_3)_4^{++}$	sp^3 tetrahedral
Hg^{++}	1.10	$Hg(NH_3)_2^{++}$	sp linear
Co^{++}	0.74	$Co(NH_3)_6^{++}$	d^2sp^3 octahedral
Co^{3+}	0.63	$Co(NH_3)_6^{3+}$	d^2sp^3 octahedral
Ni^{++}	0.72	$Ni(CN)_4^{--}$	dsp^2 square-planar
Ag^+	1.26	AgI_2^-	sp linear

There are rather few common minerals where this type of bonding is important, partly a result of the high temperature of formation of many minerals.   At elevated temperatures the coordination compounds tend to dissociate.   Nevertheless, in many processes involving mineral formation in an aqueous environment such linkages may play an important transient role.   It is also worth noting that a great number of minor elements present in the sea are held in the form of complexes, for example, $AgCl_2^-$ and $AgBr_2^-$.

Chapter 6

MIXED BONDS

In the preceding chapters we have examined some of the structures of solids, assuming that the bonding could be considered purely ionic or covalent. As with all classifications, there are numerous features of chemical compounds which suggest that the actual bonds are something in between. If the wave equation for almost any molecular system is solved, it is found that a solution

$$\psi_{\text{system}} = \psi_{\text{ionic}} + k\psi_{\text{covalent}}$$

is superior and leads to a better description of the energy and electron distribution than either covalent or ionic approximation. In the above equation, ψ_{ionic} would give the electron distribution for an ionic model, ψ_{covalent} for a covalent model, and k is a constant indicating the amount of mixing.

Let us consider a rather simple situation which might lead us to suspect that bonds have mixed character. In a solid such as ferric chloride ($FeCl_3$), the ionic model indicates that in this compound small Fe^{3+} ions are surrounded by larger Cl^- ions. Now the electron affinity of the ferric ion measured by the energy of the change $Fe^{3+} + e \rightarrow Fe^{++}$ is 43.43 ev, which is much greater than the energy of removal of the electrons from the chloride anions, 3.78 volts.

It would be anticipated that some charge transfer should occur. It may be said that the energy of interaction between the ions is sufficient to stabilize the situation, but we must not forget that when we calculate the lattice energy with the empirical Born relation, a multitude of factors are concealed in the term $1/R^n$, and we have assumed our model, not proved its existence. It must be emphasized again that the ionic model is only a *model* and is useful as long as it leads to useful correlation even if it is physically unrealistic.

Another simple case also suggests that something is wrong with our pure ionic model. Consider the structures of the compounds listed below in terms of the ionic radii.

Compound	$R_{x^{++}}$	$R_{x^{++}}/R_{s^{--}}$	Coordination number	
			Predicted	Observed
MgS	0.65	0.353	4	6
ZnS	0.74	0.402	4	4
FeS	0.76	0.413	6	6
CdS	0.97	0.527	6	4

Clearly the predictions are not very good, especially with regard to CdS and MgS, where the largest and smallest ions have precisely the wrong coordination numbers. If we were to survey a great number of such compounds, we would find that as the anion becomes more electropositive, the failure rate increases.

Many workers, and in particular Pauling, have attempted to find ways of estimating the factor k in the equation above. The argument used by Pauling is along the following lines. The ionic or covalent contribution to a mixed state leads to an energy of formation of a compound greater than that predicted on either model. If we have two molecules such as H_2 and Cl_2, the bonds may be considered fully covalent as neither atom in the homonuclear molecules can have a different affinity for the electrons in the bond. Pauling next assumed that if the HCl molecule was also fully covalent, the bond energy

$$E_{HCl} = \tfrac{1}{2}E_{H_2} + \tfrac{1}{2}E_{Cl_2}$$

In nearly all cases it is found that

$$E_{AB} > \tfrac{1}{2}E_{A_2} + \tfrac{1}{2}E_{B_2}$$

and one can write

$$E_{AB} = \tfrac{1}{2}E_{A_2} + \tfrac{1}{2}E_{B_2} + \Delta$$

The quantity Δ represents the failure to take into account the ionic contribution, and hence the magnitude of Δ should measure the amount of ionic character in the bond. Thus Δ also measures the difference in the electron affinities of A and B in the molecule A—B. As mentioned previously, the measure of an atom's tendency to gain or lose electrons is the property known as the electronegativity.

If we consider a series of compounds formed by the halogens, for example, HF, HCl, HBr, HI, inspection of the ionization potentials and electron affinities of the halogens indicates that electronegativities should decrease in the order

$$F > Cl > Br > I$$

and hence the Δ values should decrease in the same order.   Another property of the molecules also supports these considerations.   If a molecule A—B is nonpolar (fully covalent), then when the molecule is placed in an electric field, we should expect little orientation of the molecules.   If the molecule is polar or ionic, e.g.,

$$\overset{\delta+}{A}—\overset{\delta-}{B} \quad \text{or} \quad \overset{+}{A} \cdots \overset{-}{B}$$

in the field we would expect orientation.

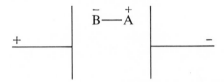

It is possible, from data on the dielectric properties of materials, to estimate the magnitude of this permanent polarization in terms of a quantity known as the dipole moment of the molecule.   The sketch below will indicate the meaning of this term.

$$\overset{\delta+}{A}\underbrace{}_{R}\overset{\delta-}{B}$$

$$\mu = \text{dipole moment} = R\delta$$

(It should be noted that only μ is measured; the value of δ thus follows from a knowledge of R and an *assumption* about the disposition of the charge.)   If the Pauling model is correct, μ as well as Δ should increase from HI to HF.   The relevant data are shown in Table 6-1.

From such data, Pauling has derived a relation between Δ and difference in electronegativities, $\Delta = 23(x_A - x_B)^2$, and has thus evaluated values for most common elements by assigning the value 2.1 to hydrogen.   The electronegativity of an element cannot be considered a precise number, but, with few exceptions, consideration of the values allows the following predictions to be made, at least qualitatively.

Table 6-1 Data Indicating Relative Polarities of Bonds in Hydrogen Halides

Bond energy	Bond energy	Δ	R_{HX}	μ_{obs}	μ if H^+—X^-	% ionic character
$H_2 = 104.2$	HF $= 134.6$	64.2	0.92	1.98	4.42	45
$F_2 = 36.6$	HCl $= 103.2$	22.1	1.28	1.03	6.07	17
$Cl_2 = 58.0$	HBr $= 87.5$	12.3	1.43	0.79	6.82	12
$Br_2 = 46.1$	HI $= 71.4$	1.2	1.62	0.38	7.74	5
$I_2 = 36.1$						

Note: R is in angstroms (A); μ is in debye units (D).

$$1\,D = \frac{\text{electronic charge} \times 1\,A}{4.774}$$

$$= \frac{4.774 \times 10^{-10} \times 10^{-8}}{4.774} = 1 \times 10^{-18} \text{ statcoulomb}$$

1. In an A—B molecule if $x_A > x_B$, the molecule will be polar in the direction

$$\overset{-}{A}—\overset{+}{B}$$

and the heat of formation from the elements will be large.

2. In a molecule A—B, if $x_A \simeq x_B$, polarity will be slight, and a covalent model will explain most of the properties.

3. If $x_A > x_B$, the ionic model will apply with greater precision as this difference becomes larger.

By assuming the simplest possible interpretation of the origin of the dipole moment as indicated above, Pauling has gone one additional stage and proposed a relation, semiquantitative, between $x_A - x_B$ and the amount of ionic or covalent character. The results are shown in Fig. 6-1. Electronegativity values for common elements are given in Table 6-2. While the exact magnitudes derived from Fig. 6-1 cannot be considered to be always reliable, trends predicted from this curve normally appear to be quite reasonable.

It is perhaps well to examine what contributions to bond character of some common compounds would be anticipated from use of Fig. 6-1. Over the range of bonds found in minerals, it cannot be expected that an ionic model, commonly most successful, will always provide reasonable answers. It is clear from the figures in Table 6-3 that a ferric ion, as such, never exists in chemical compound, as we might have expected from our earlier consideration.

The development of the Pauling electronegativity scale is rather involved, but numerous other scales and refinements have been

Table 6-2 Electronegativities of Elements

Element	Electro-negativity	Element	Electro-negativity
Hydrogen	2.1	Niobium	(1.6)
Lithium	1.0	Molybdenum[4]	(1.6)
Beryllium	1.5	Molybdenum[5]	(2.1)
Boron	2.0	Ruthenium	2.05
Carbon	2.5	Rhodium	2.1
Nitrogen	3.0	Palladium	2.0
Oxygen	3.5	Silver	1.8
Fluorine	4.0	Cadmium	1.5
Sodium	0.9	Indium	1.6
Magnesium	1.2	Tin[2]	1.65
Aluminum	1.5	Tin[4]	1.8
Silicon	1.8	Antimony[3]	1.8
Phosphorus	2.1	Antimony[5]	2.1
Sulfur	2.5	Tellurium	2.1
Chlorine	3.0	Iodine	2.6
Potassium	0.8	Cesium	0.7
Calcium	1.0	Barium	0.85
Scandium	1.3	Lanthanum	0.85
Titanium[4]	1.6	Cerium	1.05
Vanadium[3]	1.35	Praseodymium	1.1
Vanadium[4]	1.6	Hafnium	(1.3)
Vanadium[5]	1.8	Tantalum	(1.4)
Chromium[2]	1.5	Tungsten[4]	(1.6)
Chromium[3]	1.6	Tungsten[6]	2.1
Chromium[6]	(2.1)	Osmium	(2.1)
Manganese[2]	1.4	Iridium	2.1
Manganese[3]	(1.5)	Platinum	2.1
Manganese[7]	(2.3)	Gold	2.3
Iron[2]	1.65	Mercury[1]	1.8
Iron[3]	1.8	Mercury[2]	1.9
Cobalt	1.7	Thallium[1]	1.5
Nickel	1.7	Thallium[3]	1.9
Copper[1]	1.8	Lead[2]	1.6
Copper[2]	2.0	Lead[4]	1.8
Zinc	1.5	Bismuth	1.8
Gallium	1.6	Polonium	(2.0)
Germanium	1.7	85	(2.4)
Arsenic	2.0	87	0.7
Selenium	2.3	Radium	0.8
Bromine	2.8	Actinium	(1.0)
Rubidium	0.8	Thorium	1.1
Strontium	1.0	Protactinium	(1.4)
Yttrium	1.2	Uranium	1.3
Zirconium	1.4		

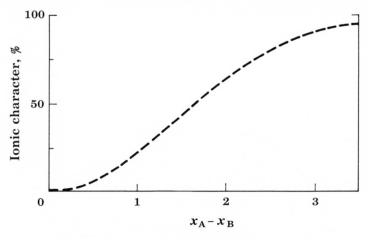

Fig. 6-1 Relationship of electronegativity difference to bond character.

proposed giving very similar values and, more significantly, similar differences in values. Perhaps one of the most obvious, and useful, scales is that due to Mulliken, who used the average of the first ionization potential and the electron affinity. If x is taken as

Table 6-3 Bond Character of Single Compounds

Compound AB	x_A	x_B	$x_A - x_B$	% ionic character
NaCl	0.9	3.0	2.1	66
KCl	0.8	3.0	2.2	70
CaCl$_2$	1.0	3.0	2.0	63
MgO	1.2	3.5	2.3	73
FeO	1.8	3.5	1.7	52
Si—O	1.8	3.5	1.7	52
Al—O	1.5	3.5	2.0	63
FeS	1.8	2.5	0.7	10
HgS	1.9	2.5	0.6	7
FeCl$_3$	1.8	3.0	1.2	30

$(I_P + E_A)/125$, the values are close to those of Pauling. While data on ionization potentials are available for most elements, electron affinity data are less common and less reliable. But as the I_P term is the largest in the sum, this quantity alone provides a very good measure of electronegativity. Ionization potentials are listed in Table 6-4.

Table 6-4 Ionization Potentials of the Elements

Element	Outer electrons	Ionization potential, ev		
		I	II	III
H	$1s^1$	13.595		
He	$1s^2$	24.580	54.40	
Li	$2s^1$	5.390	75.6193	122.420
Be	$2s^2$	9.320	18.206	153.850
B	$2s^2 2p^1$	8.296	25.149	37.920
C	$2s^2 2p^2$	11.264	24.376	47.864
N	$2s^2 2p^3$	14.54	29.605	47.426
O	$2s^2 2p^4$	13.614	35.146	54.934
F	$2s^2 2p^5$	17.42	34.98	62.646
Ne	$2s^2 2p^6$	21.559	41.07	64
Na	$3s^1$	5.138	47.29	71.65
Mg	$3s^2$	7.644	15.03	80.12
Al	$3s^2 3p^1$	5.984	18.823	28.44
Si	$3s^2 3p^2$	8.149	16.34	33.46
P	$3s^2 3p^3$	11.0	19.65	30.156
S	$3s^2 3p^4$	10.357	23.4	35.0
Cl	$3s^2 3p^5$	13.01	23.80	39.90
A	$3s^2 3p^6$	15.755	27.62	40.90
K	$4s^1$	4.339	31.81	46
Ca	$4s^2$	6.111	11.87	51.21
Sc	$3d^1 4s^2$	6.56	12.89	24.75
Ti	$3d^2 4s^2$	6.83	13.63	28.14
V	$3d^3 4s^2$	6.74	14.2	29.7
Cr	$3d^5 4s^1$	6.76	16.6	(31)
Mn	$3d^5 4s^2$	7.432	15.70	(32)
Fe	$3d^6 4s^2$	7.896	16.16	43.43
Co	$3d^7 4s^2$	7.86	17.3	
Ni	$3d^8 4s^2$	7.633	18.2	
Cu	$3d^{10} 4s^1$	7.723	20.34	29.5
Zn	$3d^{10} 4s^2$	9.391	17.89	40.0
Ga	$4s^2 4p^1$	6.00	20.43	30.6
Ge	$4s^2 4p^2$	8.13	15.86	34.07
As	$4s^2 4p^3$	10.5	20.1	28.0
Se	$4s^2 4p^4$	9.750	21.3	33.9
Br	$4s^2 4p^5$	11.84	19.1	25.7
Kr	$4s^2 4p^6$	13.996	26.4	36.8
Rb	$5s^1$	4.176	27.36	(47)
Sr	$5s^2$	5.692	10.98	
Y	$4d^1 5s^2$	6.6	12.3	20.4
Zr	$4d^2 5s^2$	6.95	13.97	24.00
Nb	$4d^4 5s^1$	6.77		24.2
Mo	$4d^5 5s^1$	7.18		
Ru	$4d^7 5s^1$	7.5		
Rh	$4d^8 5s^1$	7.7		

Table 6-4 (*Continued*)

Element	Outer electrons	Ionization potential, ev		
		I	II	III
Pd	$4d^{10}$	8.33	19.8	
Ag	$4d^{10}5s^1$	7.574	21.4	35.9
Cd	$4d^{10}5s^2$	8.991	16.84	38.0
In	$5s^25p^1$	5.785	18.79	27.9
Sn	$5s^25p^2$	7.332	14.5	30.5
Sb	$5s^25p^3$	8.64	(18)	24.7
Te	$5s^25p^4$	9.01		30.5
I	$5s^25p^5$	10.44	19.4	
Xe	$5s^25p^6$	12.127	(21.1)	32.0
Cs	$6s^1$	3.893	23.4	(35)
Ba	$6s^2$	5.210	9.95	
La	$5d^16s^2$	5.61	11.4	(20.4)
Ce	$4f^26s^2$	(6.91)	14.8	
Pr	$4f^36s^2$	(5.76)		
Nd	$4f^46s^2$	(6.31)		
Sm	$4f^66s^2$	5.6	11.4	
Eu	$4f^76s^2$	5.67	11.4	
Gd	$4f^75d^16s^2$	6.16		
Tb	$4f^96s^2$	(6.74)		
Dy	$4f^{10}6s^2$	(6.82)		
Yb	$4f^{14}6s^2$	6.2		
Lu	$4f^{14}5d^16s^2$	5.0		
Hf	$5d^26s^2$	5.5 ±	(14.8)	
Ta	$5d^36s^2$	6 ±		
W	$5d^46s^2$	7.98		
Re	$5d^56s^2$	7.87		
Os	$5d^66s^2$	8.7		
Ir	$5d^9$	9.2		
Pt	$5d^96s^1$	8.96		
Au	$5d^{10}6s^1$	9.223	19.95	
Hg	$5d^{10}6s^2$	10.434	18.65	34.3
Tl	$6s^26p^1$	6.106	20.32	29.7
Pb	$6s^26p^2$	7.415	14.96	(31.9)
Bi	$6s^26p^3$	8 ±	16.6	25.42
Rn	$6s^26p^6$	10.745		
Fr	$7s^1$			
Ra	$7s^2$	5.277	10.099	
Ac	$6d^17s^2$			
U	$5f^36d^17s^2$	4 ±		

Note: The first ionization potential represents the energy of the process

$$X \rightarrow X^+ + e$$

the second the process

$$X^+ \rightarrow X^{++} + e \quad \text{etc.}$$

Modern Trends—Universal Hybridization—the Electroneutrality Principle

It has been recognized in recent years that the quantitative application of electronegativity differences to the estimation of bond character encounters certain major difficulties, and consideration of the difficult cases has led to some important extensions of our views on chemical bonding.   Let us consider one or two of these cases.

The nitrogen trifluoride molecule NF_3 has practically no dipole moment.   Yet from consideration of $x_F - x_N = 1.0$, the bonds should have about 20% ionic character.   Further, in the structure of the NF_3 molecule, if built from p orbitals on each atom as suggested by the atomic structures

$$
\begin{array}{ccccc}
 & 1s & 2s & & 2p \\
N & (\uparrow\downarrow) & (\uparrow\downarrow) & (\uparrow)\ (\uparrow)\ (\uparrow) \\
F & (\uparrow\downarrow) & (\uparrow\downarrow) & (\uparrow\downarrow)\ (\uparrow\downarrow)\ (\uparrow)
\end{array}
$$

it would be anticipated that the

$$
\begin{array}{c}
N{-}F \\
| \\
F
\end{array}
$$

bond angles will be 90°.   But the measured bond angle is 103°, approaching a tetrahedral angle.   Both these anomalies can be readily explained if it is assumed that the $2s\ 2p$ orbitals on the nitrogen are hybridized to form something approaching an sp^3 hybrid.   Thus the molecular configuration can be considered as indicated in Fig. 6-2.   It will be seen that a lone pair of electrons in a hybrid orbital is concentrated on the side of the nitrogen atom away from the fluorine atoms.   The polarity or dipole of the molecule can now be explained as shown in Fig. 6-3.   The two contributions, by vector addition, tend to cancel, and the resultant is quite small.

A second striking case is provided by the water molecule (H_2O). As indicated previously, from the atomic structure of oxygen

$$
\begin{array}{cccc}
1s & 2s & & 2p \\
(\uparrow\downarrow) & (\uparrow\downarrow) & (\uparrow\downarrow)\ (\uparrow)\ (\uparrow)
\end{array}
$$

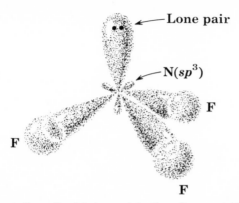

Fig. 6-2 Representation of orbitals used in the nitrogen trifluoride molecule. The nitrogen atom is in an sp^3 hybrid state with a lone pair of electrons occupying one of these hybrid orbitals.

the two bonds formed with hydrogen should be at right angles, the lone pair electrons residing in an unused p and s orbital. The observed bond angle is 104.45°, again near a tetrahedral angle, and a more realistic picture of the water molecule (Fig. 6-4) can explain many more of the features of water. Again the polarity of this molecule must be a result of terms such as those indicated in Fig. 6-5.

Even in the linear molecule HCl, a more consistent picture of the structure can be obtained if it is assumed that Cl uses an sp hybrid.

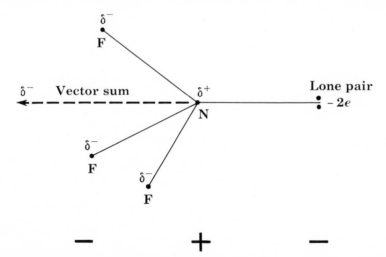

Fig. 6-3 Polarity of the NF_3 molecule. The direction of the lone-pair dipole tends to cancel the bond dipoles.

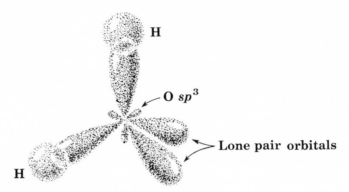

Fig. 6-4   Structure of the water molecule with two sets of lone pairs in sp^3 tetrahedral orbitals.

This hybridization leads to two large improvements:

1. Overlap and exchange interaction is increased, i.e., if in H_2O where the bond angle is 104° oxygen used p orbitals, overlap would be lost (Fig. 6-6).

2. Repulsion between unused electrons is diminished by separating these electrons into separate hybrid orbitals (Fig. 6-7).

Both these factors add to make hybridization a most common phenomenon. It is thus necessary to consider that the dipole moment is a composite term arising from lone-pair asymmetry and

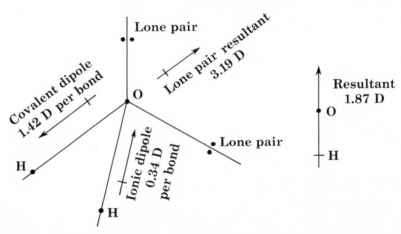

Fig. 6-5   Dipole contributions estimated by Coulson for the water molecule. Note the large lone-pair contribution and the small ionic contribution.

ionic terms.   But additional factors also enter.   Consider a molecule
such as HCl

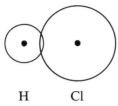

H   Cl

Even if the bonding electrons were equally shared, the unequal size
of the atoms and the unequal inner electron densities must lead to a
dipole which can be termed a pure covalent dipole.   For water, the
dipole contributions have been estimated as shown in Fig. 6-5.   It
will be noticed that the ionic term is trivial compared with other

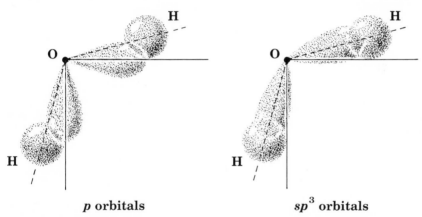

p orbitals   *sp*³ orbitals

Fig. 6-6 Structure of the water molecule illustrating the advantage of sp^3
orbitals overlapping with hydrogen at a tetrahedral angle as compared with *p*
orbitals overlapping at this angle.

contributions.   The recognition of the great significance of hybridization has largely been due to the work of Coulson and Mulliken.
From their work, it is clear that in all probability the Pauling curve
of Fig. 6-1 overestimates the ionic contribution to bonds.

Such considerations have led Pauling to propose (1948) the
electroneutrality principle.   Essentially this states that in a compound it is unlikely that the charge on any species is greater than
$\pm e$.   If the charge exceeds such a value, electrons will be transferred
to the electropositive species by covalent bond formation.   In many
ways, the success of the ionic model, in discussing structures, must

be attributed to the simple fact that the predictions are identical with those for a covalent model. When we predict bond lengths from assigned ionic or covalent radii, success may follow even though all these compounds may be dominantly covalent. We have only used different conventions in dividing a distance to obtain

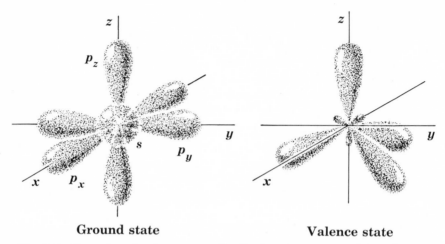

Ground state   **Valence state**

Fig. 6-7 Ground and valence state distributions in atoms such as nitrogen, oxygen, phosphorus, etc. Note that in the valence state the electrons are well separated spatially.

individual radii, but the sums of these radii must remain quite similar. As with any model, the failures are often of greater signifi- cance than the successes. We may use either model, but we should not expect precision in our predictions of bond lengths, structures, etc., when we move into bonds when $x_A - x_B$ has values in the approximate range 1 to 2.5. May we note also that whenever an atom in a molecule has an unused lone pair of electrons, this pair of electrons, stereochemically, behaves as if an additional hidden bond were present.

Chapter 7

OTHER LIMITING TYPES OF BONDS

In this chapter we shall briefly consider some types of bonds which still arise from Coulomb and exchange forces but which are normally considered separately. Several of these types lead to weaker binding forces than in the normal ionic and covalent examples, but their contributions enhance other dominant types.

Van der Waals Forces

All materials if cooled to sufficiently low temperatures form crystalline solids with significant binding energy. A case of pure van der Waals interaction is provided by the inert gases helium, neon, etc. If we consider a crystal formed from a rare gas, the bonding could only be covalent or metallic, or at least some non-polar form. But as the atom in its ground state has only paired electrons, any form of combination through covalent or metallic bonds must be initiated by unpairing or excitation, for example, He $1s^2 \leftrightharpoons$ He $1s^1 2s^1$ or He $2s^1 2p^1$. The energies of such processes are prohibitive* in the case of the inert gases and would not be compensated by bond formation, and hence the binding forces must arise in some other way. Further, when such atoms are brought into close proximity, the overlap of full electron shells must lead to strong repulsion. The binding forces are nevertheless much stronger than could be attributed to gravitational forces.

The first adequate description of such van der Waals forces was supplied by London (1930) along the following lines. The charge

---

* The recent discovery of compounds of the noble gas xenon, for example, XeF_4, indicates that with the high-atomic-number noble gases, this excitation energy is not prohibitive. As atomic numbers increase and outer orbitals are filled, the energy levels tend to crowd together.

cloud of 1*s* electrons around a helium atom is spherically symmetrical, but if we could take a photograph of a single atom with an exposure time short compared with velocity of the electrons, we would not see the statistical spherical symmetry. Thus in the short-time pic-

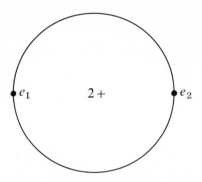

ture, the atom would be polar or dipolar. Now if two atoms are brought into proximity, there must be a tendency, arising from charge correlation, for the electron motion to become correlated or synchronized. Such a situation might be as in Fig. 7-1. We may consider that each atom induces a dipole in its neighbor, and the orientation of these induced dipoles is synchronized. Such a

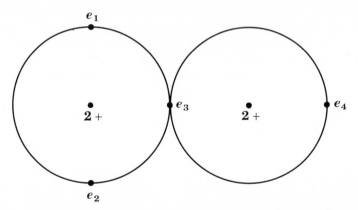

Fig. 7-1 Synchronized electron configuration in two adjacent helium atoms which could lead to van der Waals bonding.

description is adequate to explain the small binding forces which are of the order of 1000 cal or less (compare a covalent bond of 100,000 cal).

In a solid bound by such forces, clearly no directional or neutrality restrictions apply, and the most compact structure possible will be preferred. The inert gases crystallize in the cubic and hexagonal close-packed arrangements shown in Fig. 7-6, ideal structures for close packing of spheres. These structures are also common with metals.

It is again possible to speak of a van der Waals radius, easily determined by halving the internuclear distance in a crystal of the inert gas.

It should be noted that van der Waals forces operate in all solid materials and add a small contribution to the binding forces in ionic

Table 7-1   Van der Waals Radii of Atoms, A

Hydrogen	1.2	Fluorine	1.35
Nitrogen	1.5	Chlorine	1.80
Phosphorus	1.9	Bromine	1.95
Oxygen	1.40	Iodine	2.15
Sulfur	1.85		

and covalent solids. They must also play a large part in the binding of species such as solid methane (CH_4) and carbon tetrachloride (CCl_4) where the vector sum of the dipole moments in each bond is zero. In other molecules, permanent dipoles would be expected to contribute even more to the bonding energy, e.g., solid carbon monoxide, CO, CO_2, etc. Such dipole-dipole forces, enhanced by van der Waals forces, are partially responsible for the bonding of many organic molecules and the interlayers in minerals such as talc and pyrophyllite (see also page 122). It would be anticipated that the interlayer distance should be given by van der Waals radii, and this is in agreement with observation. Typical values of van der Waals radii are given in Table 7.1.

The Hydrogen Bond

There is a great deal of evidence indicating that hydrogen can form a linkage between two atoms. As the hydrogen atom, or the bare proton, is small, a maximum coordination number of 2 would be expected. But in some cases the linkage occurs between atoms with electronegativities indicating a dominantly covalent bond, and it might be concluded that hydrogen can be dicovalent. Perhaps the most striking case of such bonding is provided by salts of the type KHF_2, potassium hydrogen fluoride, where the anion HF_2^- is

a most stable species. It is clear from structural studies that this
ion is linear:

$$F\!-\!H\!-\!F$$

The immediate question raised by such a species is how can hydrogen
with a single unpaired electron form bonds with two atoms?

Let us first review some additional evidence suggesting that such
interactions occur. It is well known that water has unique properties when compared with the other hydrides of group 6—H_2S,

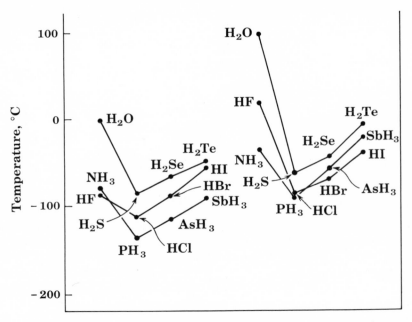

Fig. 7-2   Freezing and boiling points of groups 5, 6, and 7 hydrides.

H_2Se, H_2Te. In the same way ammonia (NH_3) differs from PH_3,
AsH_3, and SbH_3; and HF differs from HCl, HBr, HI. In Fig. 7-2
freezing and boiling points along these series are plotted. As the
molecules dissociate from the liquid, it would be expected that there
should be regular changes in relation to the molecular weight. Such
regularities exist but break down when the light members NH_3,
H_2O, HF are considered. These molecules seem to have a larger
apparent molecular weight. Anomalies are also seen in the dielectric constants (Table 7-2) where it will be noticed that molecules
containing hydrogen with O, N, or F have high dielectric constants.

The effects can be explained if it is assumed that there is strong association or interaction between the molecules. A comparison of the crystal structures of H_2O and H_2S indicates the same differences. Ice has a structure with an arrangement of oxygen atoms similar to the carbon atoms in diamond. In x-ray analysis, the hydrogen atoms are not found, but to complete the structure a hydrogen atom must be placed between each pair of oxygen atoms

Table 7-2 Dielectric Constants of Some Liquids

HF	66	HCN	123
HCl	9	NH_3	22
HBr	6	PH_3	3
HI	3	SO_2	13.8
H_2O	81.1	CO_2	1.6
H_2O_2	84	Cl_2	2.10
CH_3OH	35.4	Benzene	2.3
H_2S	10.2		

(Fig. 7-3). It cannot be said from x-ray analysis whether or not they are symmetrically placed. The structure of solid H_2S is totally different, being simply a close-packed arrangement of H_2S molecules, each sulfur being surrounded by 12 sulfur neighbors as compared with the 4 coordination of oxygens in water. The bonding in H_2S appears to be adequately explained in terms of van der Waals and dipole-dipole forces.

The minerals diaspore ($HAlO_2$) and goethite ($HFeO_2$) provide another case. In diaspore each Al atom is surrounded by an octahedral group of six oxygens and each oxygen by three aluminum atoms. If we assume an ionic structure, each oxygen must gain two electrons. Each aluminum provides $\frac{1}{2}e$ to each oxygen surrounding it. Thus the oxygen atoms gain $1\frac{1}{2}e$ from aluminum atoms. The other $\frac{1}{2}e$ must come from a hydrogen between each pair of oxygens.

We shall not go into great detail on complexities of hydrogen bonding, but whenever nitrogen, oxygen, fluorine (and possibly chlorine) atoms are present in molecules and compounds with hydrogen, there is frequently evidence to suggest that hydrogen is forming a bridge between two such atoms. The interaction is stronger than van der Waals forces but much weaker than an ionic or covalent bond (see Table 7-3). The linkage is of great structural importance whenever molecular units are bonded and is of great significance in the bonding of interlayers in clay mineral structures.

It is also responsible for effects such as the high mobility of the hydrogen ion in water where instead of the ion moving, the charge can be relayed along the hydrogen bridges.

There are two major questions regarding the hydrogen bond we should like to answer. First, are the hydrogen atoms symmetrically

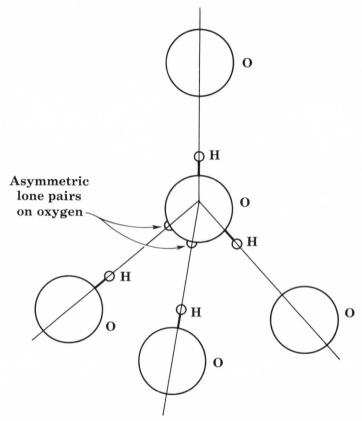

Fig. 7-3 A structural unit of ice. Note the tetrahedral arrangement of oxygen atoms and the asymmetric position of hydrogen between each pair of oxygens.

placed, and second, are the forces responsible for the bond electrostatic or exchange in origin? Present evidence suggests that in the majority of cases the hydrogen atoms are asymmetrically placed, although in some cases oscillation may not be difficult. Thus the linkage in ice must be considered to be as in Fig. 7-3. An exception to this general type is exhibited by the bifluoride ion HF_2^- where the linkage is symmetrical, but it is possible that in some oxygen species the bond is also symmetrical.

Turning to the forces in the bond, the first evidence is provided by the virtual restriction to the elements N, O, and F, all electronegative and relatively small. This suggests that polarity must be important and that dipole forces are involved. For some time a

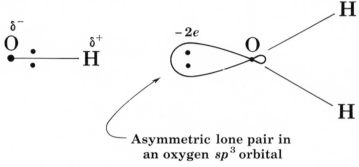

Asymmetric lone pair in
an oxygen sp^3 orbital

Fig. 7-4 Arrangement of polarity of the hydrogen bond in ice or water. The main attraction is between the proton in the OH bond and a lone pair in a hybrid orbital on a neighboring oxygen.

problem has arisen in calculating the magnitude of the hydrogen bond energy in a species such as water on the basis of dipole forces. The calculated forces were always too small. But since it has been appreciated that the dipole in molecules such as water and ammonia is largely due to the position of lone pair electrons, this difficulty has been resolved. Thus the energy of the hydrogen bond arises mainly from the interaction of the polar hydrogen in the bond with

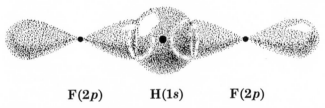

F(2p)   H(1s)   F(2p)

Fig. 7-5 Overlap of a hydrogen $1s$ orbital with two fluoride ions using p orbitals. A three-center molecular orbital is formed.

an asymmetric lone pair on O, N, or F, as in Fig. 7-4. Again, the importance of hybridization is apparent. Size is also important, for the atoms must be able to come sufficiently close to make such forces of sufficient magnitude to exert the necessary structural control.

The ion HF_2^- is probably an exceptional case. This symmetrical bridge suggests that the $1s$ orbital of the hydrogen atom may overlap equally with orbitals of both fluorines, as in Fig. 7-5. Such an effect gives rise to a three-center orbital, i.e., a molecular orbital spread over all three atoms in much the same way as the π orbitals in graphite spread over an entire planar unit in the structure. In this, the strongest hydrogen bond, exchange forces dominate. In Table 7-3 typical data on strengths and lengths of hydrogen bonds are summarized.

Table 7-3 Properties of Hydrogen Bonds

Substance	X—H—X distance, A	X—H distance covalent, A	X · · · H H bond, A	H bond energy, kcal
Ice	2.76	1.01	1.75	5.0
$(HF)_n$ gas	2.55	1.00	1.55	6
HF_2^-	2.26	0.92	1.13	58
NH_3	3.38	1.01	2.37	1.3
	N—H—F			
NH_4F	2.66	1.01	1.65	5

The Metallic Bond

To the mineralogist and petrologist, the structure of metals and the metallic bond is of less concern than all the types so far discussed, but certain features of importance in metals also appear in considerations of many minerals, particularly sulfides, tellurides, arsenides, etc., and in many ways, the properties of these are intermediate between those expected for covalent ionic and metallic compounds.

Two obvious properties are normally associated with metals: (1) a high electric conductivity, and (2) mechanical properties (ductility, malleability, etc.) unknown in ionic compounds such as NaCl and covalent compounds such as diamond.

Among the early treatments of the structure of metals was the so-called *electron gas model*. It was supposed that the atoms were ionized and the cations were packed in a medium of free electrons. Such a model would readily explain the conductivity, and the lack of any fixed directional bonds would explain the plasticity. Further, as the electron gas would occupy a small volume, the metal atoms or ions would be close-packed. The structures of most metals can be readily explained if the packing is based on a model of closest packing of spheres. Some common arrangements are shown in Fig. 7-6, and it will be seen that coordination numbers of 8 and 12 are common.

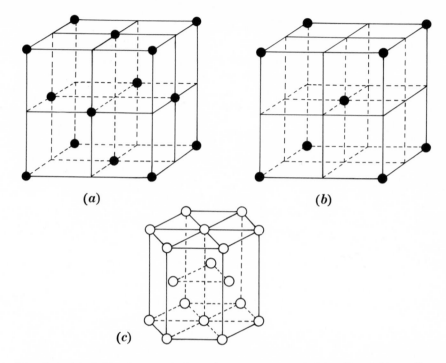

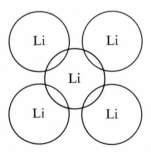

Fig. 7-6 (*a*) Cubic close-packed, each atom surrounded by 12 neighbors (for example, Cu, Ni, Fe, Ag, Au, He, Ne, A, Kr, Xe); (*b*) body-centered cubic, each atom with 8 neighbors (for example, Li, Na, K, Cr, Mo, W); (*c*) hexagonal close packing with 12 neighbors, 6 at a slightly larger distance than the other 6.

From our earlier discussion of electronegativity and bond type, the ionic model of the metal appears illogical. In a crystal of a pure metal, all atoms have the same electronegativities, and a pure covalent bond would be expected. We can approach the bonding along the following lines. Consider a crystal of lithium in which lithium is surrounded by eight near neighbors and six at a greater distance. Lithium has the structure $1s^2 2s^1$, and because promotion of the $1s$ electrons is difficult, all bonding of near neighbors must be accomplished by the single $2s$ electrons. Now this $2s$ orbital has spherical symmetry, and if we consider the atoms in a diagonal plane in the structure, we will see that all clouds interpenetrate quite symmetrically (see diagram).

We have already seen this in the F_2H^- ion and the π orbitals in graphite. The result must be the formation of an array of delocalized or multicenter orbitals extending through the entire metal crystal, just as they extend in the graphite planes. Thus if a crystal

Table 7-4   Interatomic Distances in Metals

Metal	Structure	No. of near neighbors	Distance
Aluminum	c-c-p	12	2.864
Barium	c-b-c	8	4.347
Beryllium	h-c-p	6	2.226
		6	2.286
Cadmium	h-c-p	6	2.979
		6	3.293
Calcium.......	c-c-p	12	3.947
Cesium	c-b-c	8	5.324
Chromium	c-b-c	8	2.498
Cobalt........	c-c-p	12	2.506
Copper	c-c-p	12	2.556
Gold	c-c-p	12	2.884
Iron	c-b-c	8	2.482
Lead	c-c-p	12	3.500
Lithium.......	c-b-c	8	3.039
Magnesium....	h-c-p	6	3.197
		6	3.209
Manganese	c-c-p (and others)	12	2.731
Molybdenum ..	c-b-c	8	2.725
Nickel	c-c-p	12	2.492
Potassium.....	c-b-c	8	4.544
Rubidium	c-b-c	8	4.95
Silver.........	c-c-p	12	2.889
Sodium	c-b-c	8	3.716
Tin...........	Diamond	4	2.810
	Distorted diamond	4	3.022
		2	3.181
Titanium......	h-c-p	6	2.896
		6	2.951
Zinc..........	h-c-p	6	2.665
		6	2.913

Note: c-c-p denotes cubic close packing
      c-b-c denotes body-centered cubic
      h-c-p denotes hexagonal close packing

of lithium were made of N atoms, and if they used only $2s$ orbitals, we would produce N crystal orbitals extending over the entire crystal. These would have a spectrum of energies, or a "band" of energies, and while those with the lowest energies might be full, there would be many more available orbitals and energy states empty and readily accessible. These delocalized orbitals will explain very nicely the observed properties of metals, and thus metals show to the most advanced degree the effects of crystal electron orbitals. Again for metals, we can define a metallic radius by halving the interatomic distance in the pure metal. Values are given in Table 7-4.

We may note here the distinction between an insulator like diamond and a conductor like lithium. In the former, there is an electron pair per bond, and the electrons are localized in the bonds. In the latter, there are more bonds than electron pairs, and the electrons are delocalized or smeared out through the structure. Intermediate between the two types, we have the sulfides, etc., which are semiconductors with a small number of delocalized electrons. In these, there is some conduction, but the materials are brittle, indicating the localization of most electrons. But the sulfide still retains the brilliant reflectivity of the true metal.

Cage of Clathrate Compounds

Clathrate compounds form an interesting group where compounds are formed involving no essential bonds between some parts of the solid. Their formation is almost accidental. Perhaps the most striking cases are found in the compounds of the rare gases with organic molecules such as quinol. Through hydrogen bonds, quinol forms a structure which contains rather large cavities.   These cavities are completely blocked by quinol molecules on all sides. If quinol is crystallized in the presence of a moderate partial pressure of a rare gas, the gas atoms tend to become trapped in the cavities or cages, and compounds of definite chemical composition can be formed.   The ratio of holes to quinol molecules is quite definite. Thus the compounds $3Q \cdot X$ have been formed where X is argon, krypton, xenon, acetylene, carbon dioxide, etc.   In the case of the rare gases, the only interaction with quinol will be through feeble van der Waals forces.

It is interesting to note that in potassium minerals such as ortho-clase ($KAlSi_3O_8$) decay of the K^{40} isotope to argon^{40} produces what could be called a clathrate-argon compound.   The argon in such a structure has great difficulty in escaping from its cage, and were it not for this, the decay would not lead to such a successful method of dating potassium minerals.

A case of a mineral which acts as a clathrate in the more normal sense is provided by the mineral beryl ($Be_2Al_3Si_6O_{18}$), whose struc-ture is illustrated in Fig. 9-11.   It is well known that beryl tends to accommodate helium and argon.   Inspection of the structure makes it apparent why this might happen.   The structure is built from rings of six silicon-oxygen groups.   In the center of the rings is a large cavity.   But as rings are added, they are not piled directly on top of each other, but are staggered, the staggering effectively closing each cage as it is formed.   It should be stressed that such compounds will normally be quite unstable at moderate tempera-tures; they persist only because the rate of escape of molecules from the cages is slow.

Chapter 8

A SUMMARY COMPARISON
OF LENGTHS AND STRENGTHS
OF VARIOUS BOND TYPES

In the previous chapters, limiting classes of chemical bonds have been discussed. For each bond type we can assign a given type of radius. Consideration of electronegativity differences allows some approximate estimate of what type of bond dominates in a given chemical situation and what table of radii may be expected to give the most reliable estimates. It might be anticipated that these estimates would be disappointing, but, in fact, they are often quite adequate.

It is of interest to compare some of these various radii. Frequently the term *atomic radius* is mentioned. Let us again note that the isolated atom has no definite radius; the electron cloud extends to an infinite distance from the nucleus, rapidly becoming very diffuse at distances of more than a few angstroms. The only atomic radius which has any meaning is the radius of an outer Bohr orbit or the radius of maximum probability indicated in Fig. 2-4.

In the case of all nonmetals we can choose three radii: ionic, covalent, and van der Waals, the last being the most approximate. Values of each type for chlorine, sulfur, and oxygen are listed in Table 8-1.

Table 8-1 Radii of Chlorine, Oxygen, and Sulfur

	Covalent	Ionic	Van der Waals
Chlorine	0.99	1.81	1.80
Oxygen	0.66	1.40	1.40
Sulfur	1.04	1.84	1.85

First, it is apparent that the covalent radius is much smaller than the ionic radius. This may be explained as follows. Let us assume that the radius of an atom is defined as the position of the maximum on the radial distribution curve, position A of Fig. 8-1a. When an electron is added to form the anion, the maximum will shift to

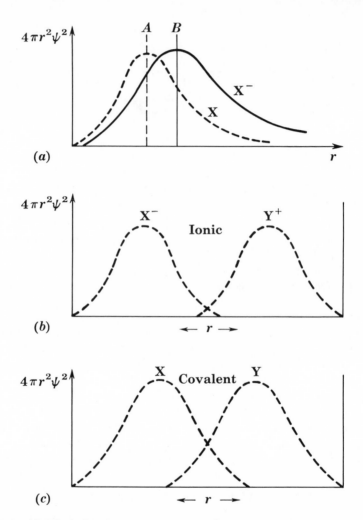

Fig. 8-1 (a) Effect of adding an electron on the outer electron radial distribution function. The maximum expands from $A \rightarrow B$. (b) Electron distribution between cation and anion in an ionic crystal. The amount of overlapping is small. (c) Electron distribution between atoms in a covalent compound with extensive penetration of outer shells.

longer distances for the number of outer electrons, and the screening has increased without any change in the nuclear charge. Thus the radial distribution curve changes to *B*. In the ionic crystal, anion and cation have full electron shells with all electrons paired. There will be little penetration of such shells (Fig. 8-1*b*). In the covalent state, the singly occupied shells overlap as much as possible, and this penetration must cause apparent shrinkage of the size of the atom (Fig. 8-1*c*).

Table 8-2   Radii of Sodium, Magnesium, and Nickel

	Ionic radius	Covalent radius*	Metallic radius
Sodium	0.95	1.59	1.57
Magnesium. . . .	0.65	1.43	1.36
Nickel	0.72	1.18	1.15

   * Values for single-bond radii taken from distances in metal hydride (MH) molecules.

The van der Waals radii are quite similar to the ionic radii, again an expected result. In the van der Waals bond, full electron shells come together, just as in the case of ionic compounds, so overlap must be small. Large distances result.

In the case of metals we have three main types of radii: ionic, metallic, and covalent. Values for sodium, magnesium, and nickel are listed in Table 8-2.

Table 8-3   Bond Length Estimation

	Sum of ions	Sum of covalent	Observed
NaCl	2.76	2.57	2.81
NaI	3.11	2.91	3.23
MgS	2.49	2.47	2.54
CuI	3.12	2.62	2.63
ZnS	2.58	2.35	2.35

Inspection of Table 8-2 makes it clear that covalent and metallic radii are quite similar, as expected for metals if the bonds are essentially covalent. The ion is smaller because in this case the outer electrons are stripped away, and those that remain are normally in an inner quantum level with a charge cloud nearer to the nucleus. The ionization process also reduces the nuclear screening.

To obtain some indication of the success of estimations consider the data in Table 8-3. It will be noted that either covalent or ionic assumption leads to a correct "order of magnitude" estimate, but the best guess is normally as expected from electronegativity considerations. Thus for NaCl the ionic sum is best, while for CuI and ZnS the ionic treatment is not. It must not be forgotten that the radii are empirical in origin, and too much must not be expected from them. As we move a given atom from one compound to another, we must expect the size to change as the different environment leads to more or less overlap of both valence and inner shell nonbonding electrons, changes in hybridization, ionization, etc.

Energies of Bonds

A comparison of bond energies can be quite misleading or ambiguous in complicated compounds. For example, in the H—Cl

Table 8-4 Dissociation Energies

Process	Type of bond	Energy, cal
$NaCl_{solid} \rightarrow Na_{gas} + Cl_{gas}$	Ionic	153,000
$CaO_{solid} \rightarrow Ca_{gas} + O_{gas}$	Ionic	257,000
$C_{diamond} \rightarrow C_{gas}$	Covalent	172,000
$HO_{gas} \rightarrow H_{gas} + O_{gas}$	Covalent	103,000
$SiC_{solid} \rightarrow Si_{gas} + C_{gas}$	Covalent	286,000
$(HF)_n \rightarrow {}_n(HF)$	H bond	7,000
$(H_2O)_{ice} \rightarrow H_2O_{gas}$	H bond	12,000
$MgSO_4 \cdot 7\,H_2O_{solid} \rightarrow MgSO_4 \cdot$ $6\,H_2O_{solid} + 1\,H_2O_{gas}$	Ion-dipole	14,000
$Ni_{solid} \rightarrow Ni_{gas}$	Metallic	101,000
$Li_{solid} \rightarrow Li_{gas}$	Metallic	37,000
$CO_{2\,solid} \rightarrow CO_{2\,gas}$	Dipole-dipole + van der Waals	6,000
$S_{solid} \rightarrow S_{gas}$	Van der Waals	3,000
$Ar_{solid} \rightarrow Ar_{gas}$	Van der Waals	1,500

gas molecule we can define the bond energy as the energy for the reaction

$$HCl_{gas} \rightarrow H_{gas} + Cl_{gas}$$

With a compound like diamond, we can talk about the energy of the process

$$C_{solid} \rightarrow C_{atom}$$

and in this case we have broken four bonds.   With NaCl, we can consider

$$NaCl_{solid} \rightarrow Na^+_{gas} + Cl^-_{gas} \qquad \text{(the lattice energy)}$$

or $$NaCl_{solid} \rightarrow Na_{gas} + Cl_{gas}$$

In each of these cases we have broken six bonds.   However, there is value in having some idea of the orders of magnitude of "energies of dissociation" of atoms in various types of compounds, even if we do not wish to use the more vague or difficult term *bond energy*. Some values are listed in Table 8-4.   All that need be said of these

Table 8-5 Force Constants of Diatomic Hydrides and Electronegativity Products (Units 10^5 dynes/cm)

	Force constant	$x_H \, x_X$	Bond dissociation energy, kcal
H—F	9.67	8.1	135
H—Cl	5.16	6.3	103
H—Br	4.12	5.88	87
H—I	3.14	5.46	71
H—O	7.76	7.35	110
H—S	4.10	5.05	81
H—N	6.03	6.3	93
H—P	3.26	4.41	76
H—C	4.37	5.25	99
H—Si	2.48	3.78	70
H—B	3.04	4.2	
H—Be	2.26	3.15	
H—Li	1.03	2.1	58

values is that the dissociation energies of covalent, ionic, and metallic compounds have rather similar magnitudes, i.e., 50 to 300 kcal. All of these bonds are difficult to break.   The H bond, dipole-dipole, and van der Waals are normally much weaker in the 1- to 10-kcal range.

Some workers have suggested that valuable information can be obtained if instead of breaking the bond between two atoms and forming atoms or ions, we simply displace the atoms from their equilibrium distance and then measure the force tending to restore the original distance.   Such a measurement can be made by studying the absorption of infrared energy which causes the atoms to vibrate. The relevant quantity is known as the *restoring force constant* or simply *force constant*.   In Table 8-5 are listed values of force constants for a wide range of hydrides.   It will be seen that quite generally the force constant increases with the electronegativity of the atom joined to hydrogen. Consideration of force constants has led to perhaps one of the best single generalizations regarding bond energies: The bond energy increases as the product of the electronegativities of the bonded atoms increases. As might be expected, there are frequent exceptions, and such a rule should be used only when measurements are not available. We are still a long way from being able to predict bond energies with the precision necessary for most geological and chemical arguments. But while absolute values may be inaccurate, relative differences may be most significant and capable of qualitative estimation.

Chapter 9

MORE COMPLEX CRYSTAL STRUCTURES

The ionic model we have discussed earlier, with due consideration of the equations for the energy of interaction of ions, leads to quite reasonable predictions of the structures of many simple crystals. The same energy considerations and relations apply to complex crystals where there are several types of ions, but now the application becomes increasingly difficult, and where several alternative structures seem possible, selection of the most probable structure may be difficult or impossible. We shall thus describe a number of more complex types and see how the simple considerations are carried over and what new considerations arise.

Just as in simple crystals consideration of the radius ratio led to prediction of coordination numbers, the same predictions may be made in complex types. Frequently, however, the numbers found may be less than predicted, but very seldom more. Many of the compounds we shall discuss in this chapter are far from being ionic, and considerable covalency in the bonds must be considered.

Oxide Structures

1. Cu_2O (Cuprite). If we write this formula as OCu_2, we might expect that a structure analogous to that of CaF_2 or TiO_2 (Figs. 4-8 and 4-9) would be possible. The ionic model would require a structure with maximum coordination of anions and cations. In this case the cuprous ion Cu^+ will coordinate 6 oxygen anions, and to maintain neutrality, oxygen would have to have 12 cuprous ions around it. Such a structure, or at least the fluorite structure, is possible but is not found. The coordination numbers are all

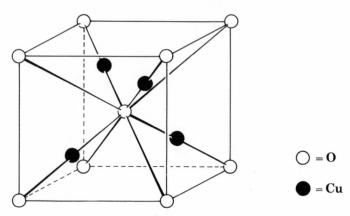

Fig. 9-1   Structure of cuprite (Cu_2O).

smaller than predicted, and each copper is linked to only two oxygen atoms and each oxygen to four cuprous ions.   The structure is shown in Fig. 9-1.   Such a structure must be explained in terms of covalent bonds.   Copper is quite electronegative, so that $x_0 - x_{Cu}$ is small, and copper may use a linear sp hybrid to form the bonds with oxygen.   The cuprous ion has a full $3d$ electron shell, and in covalent compounds a coordination number of 6 involving d^2sp^3 hybrids would not be preferred, as this would require promotion of electrons to higher energy levels.   It is interesting to note that Cu_2Se does crystallize in the fluorite structure with copper in 4 coordination.   This difference may arise because d orbitals are available with selenium, whereas oxygen is reluctant to, or does not, use its $3d$ orbitals.

2. CuO.   Unlike most transition-metal monoxides, CuO does not crystallize in the sodium chloride structure (compare TiO, VO, MnO, FeO, CoO, NiO).   In this structure each oxygen is surrounded by a tetrahedral group of four copper atoms, and each copper by a square planar group (dsp^2 hybrid bonds) of oxygen atoms. Many copper compounds show unique structures on account of extensive covalency and the full d shell of the copper atom.

3. Al_2O_3 (Corundum).   The corundum structure is typical of a number of sesquioxides including Fe_2O_3 (hematite), Ti_2O_3, Cr_2O_3. The Al—O radius ratio of 0.41 is transitional between tetrahedral and octahedral coordination.   In corundum each aluminum is surrounded by a distorted octahedral group of oxygens, and, as required by neutrality, each oxygen by a distorted tetrahedral group of aluminum atoms.   A part of the structure is shown in Fig. 9-2.

Other less stable forms of Al_2O_3 also exist and demonstrate the flexibility of possible structures. One modification crystallizes in the rock salt structure, and if we write the formula $Al_{0.66}O$, we see that this is a defect lattice with one in three cation sites vacant. Another form crystallizes in the spinel structure typical of X_3O_4 compounds. If we write the formula $Al_{2.66}O_4$, we again see that vacant cation sites will be present.

4. The Spinel Structure. Many oxides of formula X_3O_4 crystallize in the spinel structure. Some typical compounds are spinel ($MgAl_2O_4$), magnetite (Fe_3O_4), hercynite ($FeAl_2O_4$), chromite ($FeCrO_4$). All these compounds may be considered to be built from a divalent cation, two trivalent cations, and four oxygen anions. The repeating structural unit of the spinel structure contains 32 oxygen atoms or eight times the formula. In this oxygen skeleton, 8 of the possible 64 tetrahedral cation sites and 16 of the possible 32 octahedral sites are occupied. In some spinels the tetrahedral sites are preferentially taken by the divalent ions, while in other types this preference is not shown. It should be noted that in gahnite, a spinel of formula $ZnAl_2O_4$, the coordination numbers

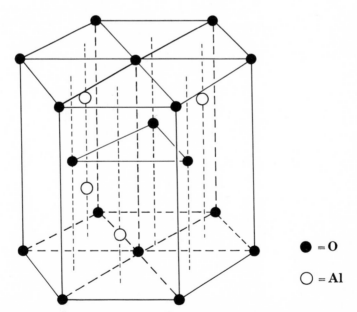

$\bullet = O$

$\bigcirc = Al$

Fig. 9-2 Part of the corundum structure. Note that the oxygen atoms are in hexagonal close packing (see Fig. 7-6), and that the structure essentially shows layers of O and Al, with one-third of the possible Al sites vacant.

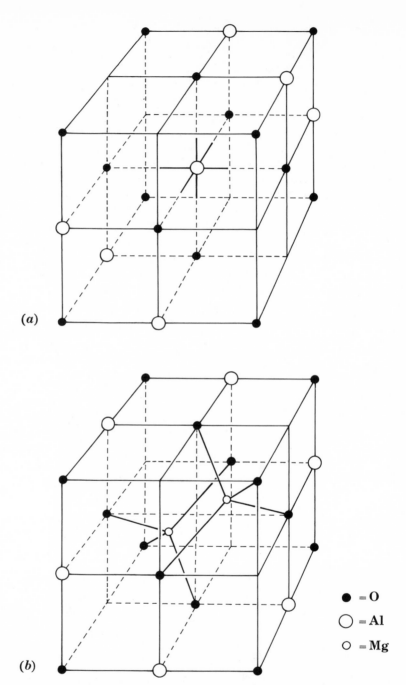

Fig. 9-3 The spinel structure. In this structure the oxygen atoms are in cubic close packing (see Fig. 7-6). The entire structure is built by joining the two types of unit shown. Unit *a* contains the metal ion in octahedral coordination, while unit *b* contains the metal ions in tetrahedral coordination. Note the vacant sites.

of $Zn^{++}(4)$ and $Al^{3+}(6)$ are in opposition to the arrangement that would be expected from size considerations, and possibly reflect the empty $3d$ shell of aluminum and the full $3d$ shell of zinc. Many complex factors must operate in such selections. The structure is illustrated in Fig. 9-3.

Sulfides

A great number of common sulfides crystallize in the halite and sphalerite structures (Figs. 4-5 and 4-7 and Tables 4-3 and 4-5).

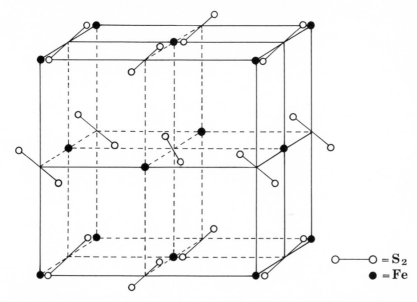

$\circ\!\!-\!\!-\!\!-\!\!\circ = S_2$

$\bullet = Fe$

Fig. 9-4 The pyrite (FeS_2) structure containing the dumbbell-shaped S_2^{--} anion.

These substances are now dominantly covalent, and the coordinate structures will reflect hybrid orbitals more than radius ratios. We shall mention only two additional structures, both of disulfides.

1. Pyrite Type. Sulfides, such as FeS_2 (pyrite), MnS_2, CoS_2, NiS_2, are all compounds containing divalent cations. It would be erroneous to consider them to be built from formal oxidation states $X^{4+} + 2S^{--}$, as sulfur is a reducing species. The pyrite structure thus contains the S_2^{--} anion. The structure is illustrated in Fig. 9-4, and a glance at this figure will indicate that we are dealing with a modified sodium chloride structure, modified on account of the asymmetry of the anion.

2. Molybdenite (MoS$_2$). Molybdenite is characterized by extreme softness and excellent cleavage in one direction. Recently it has come into wide use as a lubricant in many ways superior to graphite.

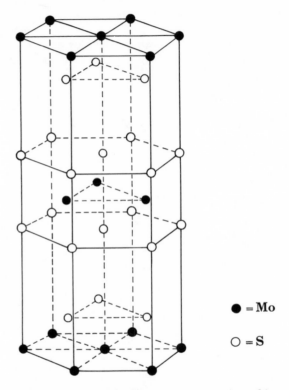

Fig. 9-5 Structure of molybdenite (MoS$_2$). The structure consists of layers of molybdenum and sulfur atoms, but the order of layers is ——S—Mo—S—— S—Mo—S——. A weak linkage occurs between the S——S layers.

One would certainly anticipate some structure related to graphite. The structure is shown in Fig. 9-5. It will be noticed that the structure contains layers of molybdenum and sulfur atoms, and our concept of achieving maximum stability of the ionic crystal by keeping like species as far apart as possible has broken down completely. Each molybdenum is surrounded by six sulfur atoms, and each sulfur atom is linked to three molybdenum atoms. Each layer in the crystal consists of a giant two-dimensional sheet or molecule of formula $(MoS_2)_n$. Between the sheets, opposing layers of sulfur atoms face each other and will be linked by van der Waals forces. Such a structure would not be expected if the ions carried full formal charges.

Halides

We have already seen that a large number of simple halides of types AB and AB_2 crystallize in the NaCl, CsCl, CaF_2, and TiO_2 structures (Figs. 4-5, 4-6, 4-8, and 4-9). However, many chlorides, bromides, and iodides are sufficiently covalent that other structures and bond arrangements occur. Let us consider a few typical types.

Many transition-metal dihalides, and even some hydroxides, crystallize in the CdI_2 and $CdCl_2$ layer lattice structures. In the structures of CdI_2 and $CdCl_2$ the halogen atoms are in hexagonal and cubic close packing, respectively (see Fig. 7-6), with metal atoms filling some of the interstices in these arrangements. The difference between the layer lattice and normal ionic lattice is shown schematically in Fig. 9-6. In each structure the halogen is linked to two metal ions, and each metal ion to four halogens. In the two-dimensional case, the layer lattice (a) consists of infinite chainlike molecules, while the ionic lattice (b) forms an infinite sheet.

The structure of $CdCl_2$ is quite similar to that of MoS_2 (Fig. 9-5). These layer structures tend to be preferred by substances with dominantly covalent bonds, and the weak interlayer forces are essentially van der Waals and dipole in type. The perfect cleavage between the layers is as expected.

Many trihalides such as $FeCl_3$ and $CrCl_3$ occur in similar layer structures. Another type is illustrated by aluminum chloride $(AlCl_3)$. It is well known that in the vapor phase $AlCl_3$ forms the dimer molecule Al_2Cl_6. We can draw a structure for this molecule as indicated below, where each Al is tetrahedrally surrounded by

four chlorine atoms. Two chlorine atoms form a bridge binding the dimer together by the use of a lone pair of electrons which interacts with an empty sp^3 orbital on the aluminum. Solid Al_2Cl_6 also

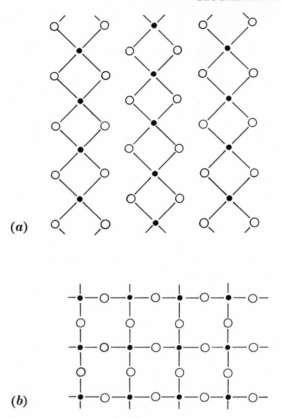

Fig. 9-6 Two-dimensional analogues of a layer lattice (*a*) and the normal ionic lattice (*b*).

forms a layer lattice but is probably best described in terms of close packing of these Al_2Cl_6 molecules.

Hydroxides

Few simple hydroxides crystallize in the NaCl and CaF_2 structures, but many occur in less symmetrical modifications or layer lattices. The hydroxyl ion, while similar in size to fluorine or oxygen, cannot be considered spherically symmetrical. Further, hydrogen bonds of moderate strength may influence structures and, in particular, cause stronger interlayer linkage in layer lattices. Simple hydroxides such as KOH crystallize in a tetragonal modification of the NaCl structure. This particular hydroxide becomes cubic with the NaCl structure on heating, presumably because the thermal agitation allows free rotation of the OH^- ion and produces a pseudospherical symmetry of the anion. Gibbsite ($Al(OH)_3$) forms a typical layer lattice with hydrogen bonds joining layers. Brucite ($Mg(OH)_2$) forms a cadmium iodide layer lattice. Diaspore ($HAlO_2$) and goethite ($HFeO_2$) also form layer-type lattices with hydrogen bonds playing a large part in the structure.

Oxyanion Structures

With compounds such as nitrates, sulfates, carbonates, and silicates, because so much of their reaction chemistry is concerned with identity of an oxyanion, we may consider them from such a viewpoint initially. With the simplest approach we might assign a radius to the entire oxyanion and then apply the simple arguments of Chap. 4. Such an approach is bound to be inadequate because of lack of symmetry of the anion.

1. Potassium Perchlorate ($KClO_4$). At elevated temperatures, potassium perchlorate crystallizes in the sodium chloride structure but at lower temperatures in a layer lattice similar to anhydrite ($CaSO_4$). We may consider the tetrahedral perchlorate ion to be made from a Cl^{7+} ion of 0.26 A radius, and four oxygen ions (O^{2-}) of radius 1.40 A. The effective radius of the whole complex is thus about 3.06 A. Clearly this large anion could take a large number of potassium ions around it, but as this is a 1:1 compound, neutrality requires that the number of ClO_4^- around potassium control the structure. The radius ratio $K^+/ClO_4^- = 0.43$ will allow 6:6 coordination and the structure (Fig. 9-7) is quite analogous to the NaCl structure.

2. The Calcite and Aragonite Structures. Many carbonates of formula $M^{++}CO_3^{--}$ and nitrates ($M^+NO_3^-$) crystallize in these two structures. The nitrate and carbonate groups, as anticipated from the electronic structure and size of the atoms, are 3-coordinated and form trigonal-planar anions. Clearly no simple structure is to be anticipated, as this group is far from being spherically symmetrical. The calcite structure is illustrated in Fig. 9-8. In this structure each calcium ion is surrounded by six oxygen atoms. Now the calcite group carbonates, which include $CaCO_3$, $MgCO_3$, $FeCO_3$, $MnCO_3$, $ZnCO_3$, and the nitrates $NaNO_3$, KNO_3, all crystallize in rhombohedrons which can be likened to a distorted cube. The calcite structure can be derived from the sodium chloride structure by taking the NaCl unit (Fig. 4-5) and compressing along a diagonal of the cube. It may be noted that carbonates in this structure have cations with radii of 1 A or less and thus could not coordinate more than six oxygens from a consideration of radius ratios.

The carbonates of larger cations now crystallize in the aragonite structure ($CaCO_3$, $SrCO_3$, $BaCO_3$, KNO_3). In this structure which still contains the planar carbonate or nitrate ion, each cation is surrounded by nine oxygen atoms again demonstrating the

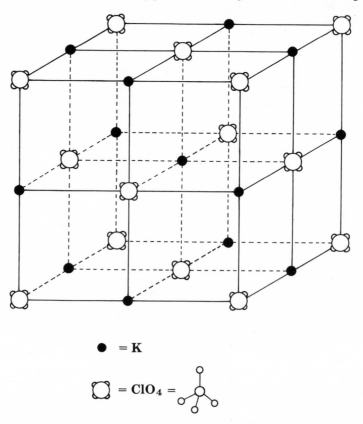

$\bullet$ = K

$\square$ = ClO_4 =

Fig. 9-7 Structure of $KClO_4$ (potassium perchlorate). Note the similarity to the sodium chloride structure (Fig. 4-5).

tendency to obtain a maximum number of neighbors of opposite polarity.

As discussed in Chap. 11, several of these substances crystallize in both structures depending on pressure and temperature. The aragonite structure is favored by increasing pressure, and the calcite structure by increasing temperature. Such transitions are shown by $CaCO_3$ and KNO_3.

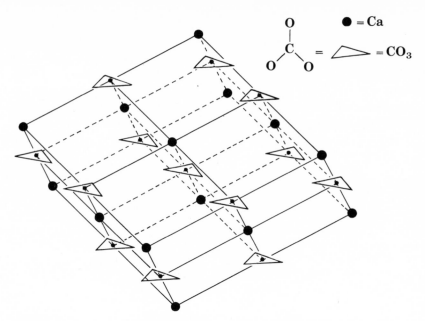

Fig. 9-8 Structure of calcite ($CaCO_3$). Note the distorted sodium chloride structure.

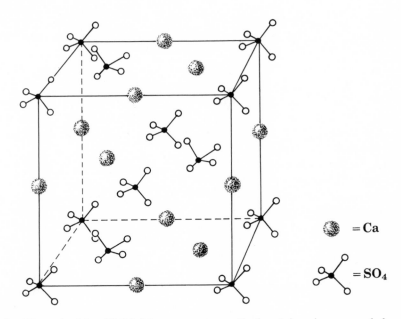

Fig. 9-9 The anhydrite ($CaSO_4$) layer structure. Each calcium is surrounded by eight oxygens.

3. The Anhydrite Structure.   Calcium sulfate and potassium per-chlorate (at low temperatures) crystallize in a layer lattice.   The structure is indicated in Fig. 9-9. Each sulfur is tetrahedrally sur-rounded by oxygens, and each calcium linked to eight oxygens.

4. The Perovskite Structure.   This structure is formed by many salts where the anionic complex has a coordination number of 6 and forms an $[XY_6]$-type anion.   Typical species are perovskite ($CaTiO_3$),

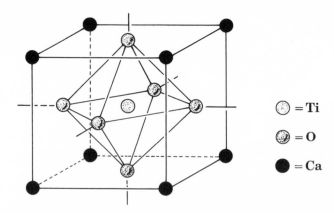

Fig. 9-10   The structure of perovskite ($CaTiO_3$).

potassium iodate (KIO_3), and potassium nickel fluoride ($KNiF_3$). It may well be asked why some simple structure based on an XY_3 anion is not found as with nitrates and carbonates. In part the answer must be that this would not allow maximum coordination, for in each of the above cases the I, Ni, or Ti can take six oxygen or fluorine neighbors.

The perovskite structure is illustrated in Fig. 9-10.   At first sight the structure shows some resemblance to the cesium chloride struc-ture with one major difference.   Each oxygen of the TiO_6^{8-} group is shared between two titanium atoms, the Ti—O—Ti bridge being linear.   We may thus consider the structure to be formed from an array of TiO_6 groups linked to each other by the corners.   Above the center of each face of the octahedral group a calcium atom is found.

On an ionic model this structure would be considered to be formed from Ca^{++}, Ti^{4+}, $3O^{--}$.   It is interesting to see how these charge relations, and the resulting neutrality, are obtained.   As each titanium atom loses four electrons, it will give $\frac{4}{6}e$ to each of the six oxygen atoms surrounding it. As each oxygen is linked to two titanium atoms, from titanium neighbors it receives $\frac{8}{6}e$. From the structure it will be seen that each oxygen is surrounded by four calcium atoms so must receive the remaining $\frac{4}{6}e$ from these, or $\frac{1}{6}e$ from each. The coordination number of calcium follows, for each calcium loses two electrons and hence must have 12 oxygen neighbors. This device of satisfying formal charge requirements can be most useful in analyzing a possible structure. Further it is normally found that this charge satisfaction must involve near neighbors only, and any structure where longer-range transfer must be considered is not favored. In other words, large structural units must not have residual charge, or electrons paired in covalent bonds are not transferred far from their origin.

Silicates

The silicates form a group of inorganic compounds of great chemical complexity, far greater than the oxymetal compounds of sulfur, nitrogen, or carbon. Until the introduction of x-ray diffraction techniques with the possibility of studying atomic arrangements in solids, the principles underlying silicate structures were beyond the reach of the chemist, partly because the species in the solids have no counterparts in solution. The mineralogist and crystallographer who have studied the external form which reflects the internal arrangement have made far more progress. It is clear, for example, that two silicates which possess a micaceous cleavage might be expected to have some structural feature in common. But chemically, such analogies might not be apparent; for example, compare the formulas of muscovite, a mica ($H_2KAl_3Si_3O_{12}$), and the mineral kaolinite $H_4Al_2Si_2O_9$.

The general features of silicate structures become reasonable through the following considerations:

1. The silicon atom is always surrounded by a tetrahedral group of four oxygen atoms, as would be expected from radius ratio considerations or use of an sp^3 hybrid on the silicon atom forming covalent bonds.

2. These silicate ions may occur as separate units or may be polymerized to varying degrees by sharing a corner oxygen.

3. As the aluminum ion (Al^{3+}) is similar in size to the silicon ion (Si^{4+}), mutual replacement is possible. This replacement seldom exceeds 50%.

We have already seen the sharing of an oxygen on an oxyanion in the perovskite structure.   When SiO_4 groups polymerize, the corner atoms are shared but never edges or sides.   Preference for such corner sharing or linear sharing can be explained in terms of obtaining minimum cation-cation repulsion, e.g.,

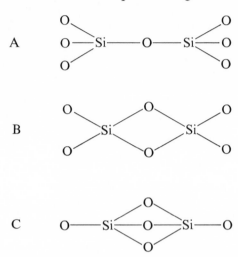

In both the latter structures, B and C, the cation repulsion must be larger than in the first.   But the story is more complex than this. In the linear structure, oxygen will use an *sp* linear hybrid, and *sp* hybrids form stronger bonds than any bent hybrid such as sp^3. The linear *sp* oxygen also brings *p* orbitals on the oxygen into a position suitable for π bond formation with empty $3d$ orbitals on the silicon.   It is well known that the Si—O bond is rather shorter than expected, which may reflect this $d\pi$ bonding.

If the stoichiometry of a compound permits, polymerization does not occur, and isolated units form which we can consider to contain the anion SiO_4^{4-}.   Typical examples are provided by the olivine group

$$Mg_2SiO_4 \qquad \text{Forsterite}$$

$$Fe_2SiO_4 \qquad \text{Fayalite}$$

the garnet group

$$Ca_3Al_2(SiO_4)_3 \qquad \text{Grossularite}$$

$$Mg_3Al_2(SiO_4)_3 \qquad \text{Pyrope}$$

$$Fe_3Al_2(SiO_4)_3 \qquad \text{Almandine} \qquad \text{etc.}$$

and minerals such as

$ZrSiO_4$ Zircon

Be_2SiO_4 Phenacite

Zn_2SiO_4 Willemite

With all these minerals, no oxygen is linked to two silicon atoms, and the tetrahedrons are isolated. Zircon ($ZrSiO_4$) has a structure quite analogous to anhydrite ($CaSO_4$), already illustrated.

In the garnet family of formula $X_3Y_2(SiO_4)_3$, each X atom is in 8 coordination with oxygen, and each Y atom in 6 coordination. Pyrope ($Mg_3Al_2(SiO_4)_3$) is one of the few compounds containing the magnesium ion in 8 coordination. The phase is dense and characteristic of high-pressure environments.

Dimers

The most simple polymerization of the SiO_4^{4-} group is the formation of a dimer:

Within the dimer, as always, the silicon is in tetrahedral coordination with oxygen. The Si—O—Si bridge is seldom strictly linear. Examples of such silicates are provided by

$Sc_2Si_2O_7$ Thortveitite

$Zn_4Si_2O_7(OH)_2 \cdot H_2O$ Hemimorphite

$Ca_2MgSi_2O_7$ Melilite

$CaAl_2Si_2O_7(OH)_2 \cdot H_2O$ Lawsonite

Rings

Polymerization does not now proceed in single units, but from the dimers we pass to rings and infinite chains. The common rings contain the units $(Si_3O_9)^{6-}$ as in the mineral benitoite ($BaTiSi_3O_9$); $(Si_4O_{12})^{8-}$, as in the complex mineral axinite; and $(Si_6O_{18})^{12-}$, as in

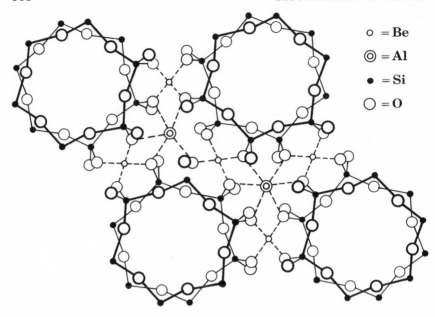

o = Be

◎ = Al

• = Si

◯ = O

Fig. 9-11   The ring structure of beryl ($Be_3Al_2Si_6O_{18}$), looking down on the rings. Note that superposed rings are displaced.

the mineral beryl ($Be_3Al_2Si_6O_{18}$). As might be expected, the 3 and 6 rings are reflected in the external threefold and sixfold symmetry of these crystals. The structure of beryl is illustrated in Fig. 9-11.

Chains

As complexity increases, any concept of a finite anion in the structure disappears. With chain polymerization the anion becomes $(SiO_3^{--})_n$, where n is a very large number. It would be expected that the presence of an anionic chain would be reflected in exterior form, and silicates with such a structure do form elongated crystals, particularly when synthesized. In this structural group is the large mineral family of pyroxenes. Examples are

$MgSiO_3$	Enstatite
$CaMgSi_2O_6$	Diopside
$NaFeSi_2O_6$	Acmite
$NaAlSi_2O_6$	Jadeite
$LiAlSi_2O_6$	Spodumene

The various silicate polymer types are shown in Fig. 9-12, while the structure of diopside is shown in Fig. 9-13.

By cross linkage of two chains (Fig. 9-12) a band is formed with the repeat unit being $[Si_4O_{11}]^{6-}$. Again the band is responsible for the production of elongated crystals, and some forms of asbestos crystallize with this structure. It is noteworthy that the cross-linked band appears to be structurally stronger than the chain, and accentuated linear crystal habit is more pronounced with amphiboles than pyroxenes. The structure of the amphibole tremolite is shown in Fig. 9-14. In both amphiboles and pyroxenes the chains run parallel to the elongation of the crystal. When a crystal is viewed in end section normal to the chains, the difference in the chain and band dimensions is reflected in the cleavages shown (Fig. 9-15). Typical members of the amphibole family are listed below.

$Mg_7Si_8O_{22}(OH)_2$	Anthophyllite
$Ca_2Mg_5Si_8O_{22}(OH)_2$	Tremolite
$Fe_7Si_8O_{22}(OH)_2$	Grunerite
$Na_2Fe_2{}^{3+}Fe_3{}^{++}Si_8O_{22}(OH)_2$	Riebeckite

It will be seen that the formula can be written generally as A_2B_5, $(SiAl)_8O_{22}(OH)_2$ where the A ions may be Ca^{++}, Mg^{++}, Fe^{++}, Na^+; the B atoms Mg^{++}, Fe^{++}, Fe^{3+}, Al^{3+}, Ti^{3+}, etc. As well, the hydroxyl ion may be replaced in part by the fluoride ion and oxygen. The common amphibole of igneous rocks, hornblende, has a most complex chemistry.

The mineral chrysotile $((OH)_6Mg_6Si_4O_{11}\cdot H_2O)$, commonly used as a form of asbestos, has a more complex structure combining the silicate band structure and brucite layer structure (cf. the structure of chlorite).

Sheet Structures

Clearly if bands are further polymerized sideways, an anion in the form of an infinite two-dimensional sheet with repeat unit $(Si_4O_{10})^{4-}$ is formed. This structure (Fig. 9-12) is responsible for the great structural similarities of the minerals of the clay, mica, and chlorite groups. All show a platy habit with a perfect direction of easy splitting parallel to the sheets. The ease of such splitting is controlled by the interlayer forces, in some cases ionic in nature, in

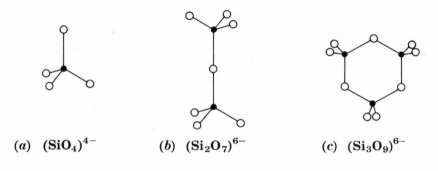

(a) $(SiO_4)^{4-}$ (b) $(Si_2O_7)^{6-}$ (c) $(Si_3O_9)^{6-}$

(d) $(Si_6O_{18})^{12-}$

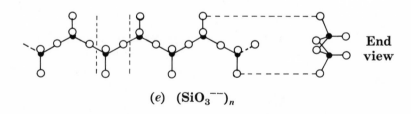

(e) $(SiO_3^{--})_n$

Fig. 9-12 Silicate groupings: (a) isolated unit; (b) dimer; (c, d) rings; (e) infinite chain; (f) infinite band; (g) infinite sheet. (*Continued on p.* 121.)

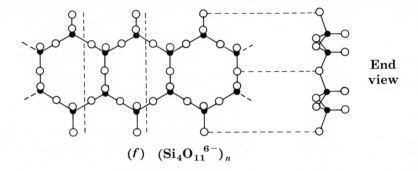

End
view

(f) $(Si_4O_{11}^{6-})_n$

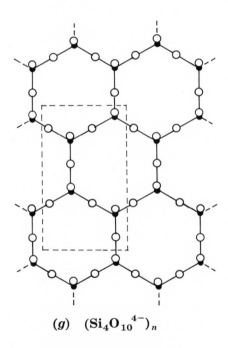

(g) $(Si_4O_{10}^{4-})_n$

Fig. 9-12 *(Continued)*

Note that the formal charge on the repeat unit can be obtained by adding the number of oxygen atoms joined to one silicon only.

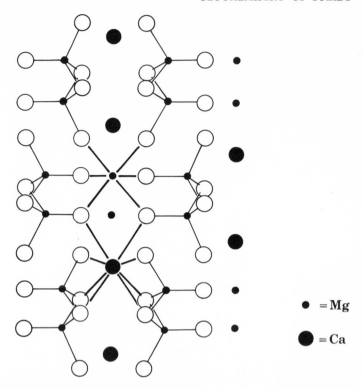

Fig. 9-13 Simplified view of the diopside ($CaMgSi_2O_6$) structure. The section is taken normal to the chains and shows the end view (cf. Fig. 9-12*e*). Two types of cation sites are present, one having 8 coordination and one 6 coordination. For this reason any pyroxene formula can be written $XYSi_2O_6$. The sites of 8 coordination are preferentially occupied by larger cations, in this case Ca^{++}.

others hydrogen bonds and van der Waals forces. The structure of the mica muscovite is shown in Fig. 9-16 and is schematically compared with some other layer types in Fig. 9-17. Recent studies of sheet structure silicates have shown a great range of possible complexities in that the layers typical of several end members may be interleaved.

Three-dimensional Polymers

In the final stage of polymerization of the SiO_4 tetrahedrons, each oxygen is shared between two silicon atoms. Clearly such a framework is electrically neutral and more complex compounds can only

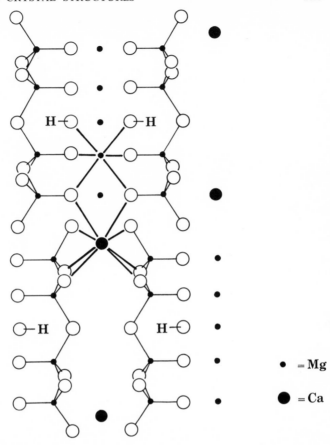

Fig. 9-14 Simplified view of actinolite ($Ca_2Mg_5Si_8O_{22}(OH)_2$). The section is normal to the bands and shows the end view (Fig. 9-12*f*). Again two types of cation sites are present in the ratio 2:5, so the amphibole formula can be written $X_2Y_5Si_8O_{22}(OH)_2$. Large cations occupy the 8-coordinated X sites. The hydroxyl ions occupy positions in the center of the rings of the band (Fig. 9-12), positions not available in a pyroxene. Note the weak linkage between bands which are back to back.

form by the introduction of ions of different valence to replace some of the silicon atoms. Thus we may pass from

$$SiO_2 \rightarrow K^+ (AlSi_3O_8)^-$$
Orthoclase

or to $$CaAl_2Si_2O_8$$
Anorthite

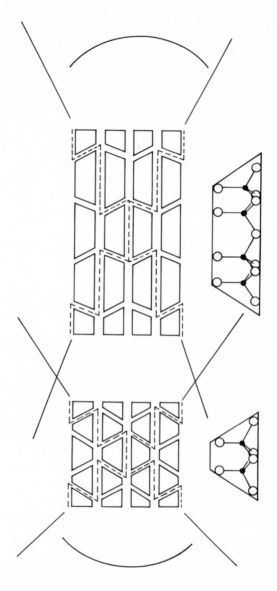

Fig. 9-15 Explanation of the more acute cleavage angle found in amphiboles as compared with pyroxenes in sections normal to chains and bands. The breaking is shown to occur along the weak back-to-back position.

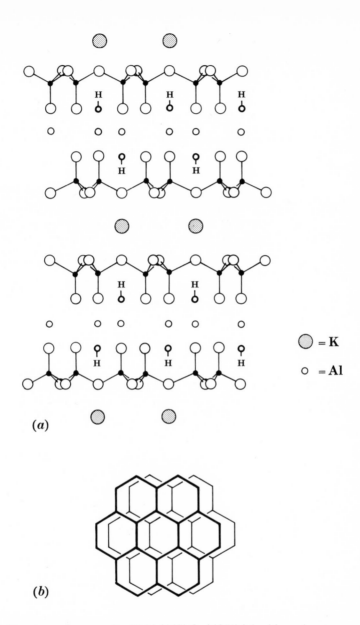

$\bigcirc$ = K

$\circ$ = Al

(a)

(b)

Fig. 9-16 (a) Structure of muscovite $(KAl_2(AlSi_3O_{10})(OH)_2)$ looking edgeways at the sheet. The structure consists of a double sandwich of alumino silicate sheets joined by aluminum. This unit has the basic formula $[Al_2AlSi_3O_{10}(OH)_2]^-$, and potassium ions join the layers together. Again, hydroxyl ions occupy the positions in the center of silicon oxygen rings (Fig. 9-12). (b) View looking down on sheets to show staggering of the rings.

125

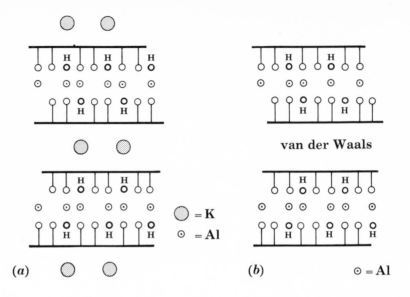

van der Waals

◯ = K
⊙ = Al

(a)　　　　　　　　　　　(b)　　　　　⊙ = Al

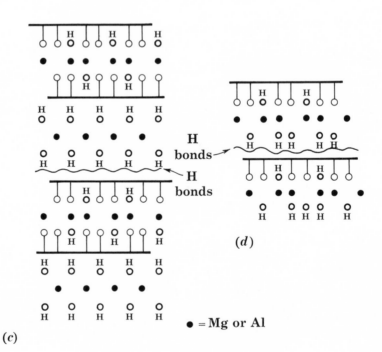

H bonds

H bonds

(d)

● = Mg or Al

(c)

Fig. 9-17 See caption on facing page.

Many important classes of silicates crystallize in such states of three-dimensional polymerization. These include:

1. The silica family, quartz, tridymite, cristobalite, coesite, etc.
2. The feldspars, e.g.,

$KAlSi_3O_8$	Orthoclase
$NaAlSi_3O_8$	Albite
$CaAlSi_2O_8$	Anorthite
$(NaCa)_{1-2}Al_{1-2}Si_{2-3}O_8$	Plagioclase, a solid solution series at high temperatures

3. The feldspathoids, e.g.,

$NaAlSiO_4$	Nepheline
$KAlSiO_4$	Kaliophilite
$KAlSi_2O_6$	Leucite

4. The zeolites, e.g.,

$NaAlSi_2O_6 \cdot H_2O$	Analcime
$Na_2(Al_2Si_3O_{10}) \cdot 2H_2O$	Natrolite
$CaAl_2Si_4O_{12} \cdot 4H_2O$	Laumontite

and a host of others.

Fig. 9-17 A schematic comparison of some sheet structure silicates. Note that the silicon oxygen sheet is shown as a line with active oxygens pointing down.

(a) Muscovite (cf. Fig. 9-16a). The brittle micas, for example, margarite $(CaAl_2(Al_2Si_2O_{10})(OH)_2)$, have identical structures, but now the sheet "anion" is $[Al_2Al_2Si_2O_{10}(OH)_2]^{2-}$ linked by Ca^{++} in place of $[Al_2AlSi_3O_{10}(OH)_2]^-$ linked by K^+ as in muscovite. The interlayer binding forces are hence about four times as strong.

(b) Pyrophyllite $(Al_2(Si_4O_{10})(OH)_2)$ has a mica structure. As the sheet unit $Al_2Si_4O_{10}(OH)_2$ is neutral, no interlayer ions are needed. The double sheets are linked only by van der Waals forces, and hence cleavage is perfect and the material soft. Talc $(Mg_3Si_4O_{10}(OH)_2)$ has the same structure.

(c) Chlorite $(AlMg_5(AlSi_3O_{10})(OH)_8)$. The structure is now composed of interleaved pyrophyllite and brucite $(Mg(OH)_2)$ layers. Brucite forms a layer lattice. The unit formulas of the two layers approach:

$$(Mg_3(AlSi_3O_{10})(OH)_2)^- \quad \text{and} \quad (Mg_2Al(OH)_6)^+$$

forming two large ions. The interlayer binding forces are very weak.

(d) Kaolinite $(Al_4Si_4O_{10}(OH)_8)$. In this structure there is no silicate sheet sandwich. A single sheet is followed by a cation and hydroxyl ion layer. The units are joined by hydrogen bonds. The unit is weak (compare a steel bridge girder) and buckles, and in some clays even rolls into tubes. Water and other hydrogen-bonding materials may enter between the layers and cause swelling.

It is quite beyond the scope of this book to consider the structural complexities and polymorphism of this group of structures. The structure of β-cristobalite is shown in Fig. 9-18. The zeolites present some most interesting features. In their structures the silicon-aluminum tetrahedrons are joined to form an open framework which in some cases contains channels several angstroms wide. These channels provide easy access to the interior of the crystal and also accommodate the water molecules. Further, many zeolites

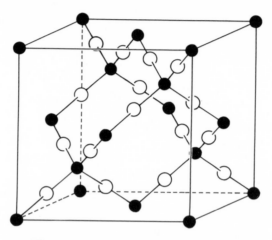

Fig. 9-18   Structure of β-cristobalite.

contain many more potential cation sites than cations. This set of structural features leads to the phenomena of easy cation exchange or base exchange (see page 136) and to molecular sieve action.

With some zeolites, mild heating will remove the water from the channels without collapse of the framework. The dehydrated crystal is now in a state to take other molecules if their polarity and geometry allow entry.   The zeolite acts as a molecular sieve.   Molecules such as ethyl alcohol (C_2H_5OH) and iodine (I_2) enter with ease. But if two molecules such as

(a)

$$H-\overset{\overset{\displaystyle H}{|}}{\underset{\underset{\displaystyle H}{|}}{C}}-\overset{\overset{\displaystyle H}{|}}{\underset{\underset{\displaystyle H}{|}}{C}}-\overset{\overset{\displaystyle H}{|}}{\underset{\underset{\displaystyle H}{|}}{C}}-\overset{\overset{\displaystyle H}{|}}{\underset{\underset{\displaystyle H}{|}}{C}}-\overset{\overset{\displaystyle H}{|}}{\underset{\underset{\displaystyle H}{|}}{C}}-\overset{\overset{\displaystyle H}{|}}{\underset{\underset{\displaystyle H}{|}}{C}}-H$$

and

(b)

approach, the molecule with the branched chain may be excluded while the straight chain is accepted. Further, as the cations in the zeolite structure are located on the sides of the channels, by exchanging cations of different sizes the channel dimensions and sieve sizes can be selectively changed.

The general features of silicate structures outlined above have reduced the chemically complex family to a rational and quite simple set of structural types. As studies proceed, more second-order complexities are encountered—for example, the mixed-layer sheet structures and minerals with more than one structural unit present. Nevertheless, the structural approach was successful in unraveling this problem where classical chemical approaches failed.

Other Polymer Types

Polymerization of the silicate type is by no means restricted to silicates; borates, germanates, tungstates, phosphates, etc., may show similar structures. The borates are perhaps worth considering further, as the similarities are considerable, with a major difference in that the boron atom frequently accommodates only three oxygen atoms in a group which is trigonal-planar in configuration. Also, a tetrahedral grouping is known analogous to that in silicates.

As with the silicates, structures with isolated BO_3^{3-} groups, dimers ($B_2O_5^{4-}$ groups), chains and rings ($B_3O_6^{3-}$) are found. Possible polymer types are indicated in Fig. 9-19.

Conclusion

In conclusion we may repeat certain arguments which are useful in the consideration of complex structures.

1. In ionic structures, the number of anions around a cation can be estimated from radius ratio considerations. As structures become more covalent, coordination is determined by available hybrid

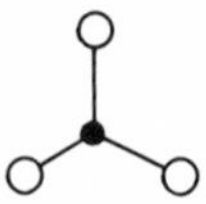

BO_3^{3-}

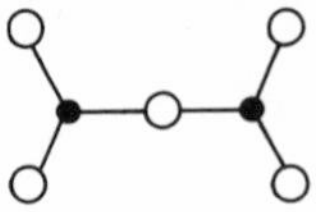

$B_2O_5^{4-}$

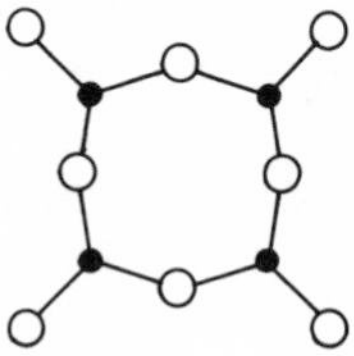

$B_4O_8^{4-}$

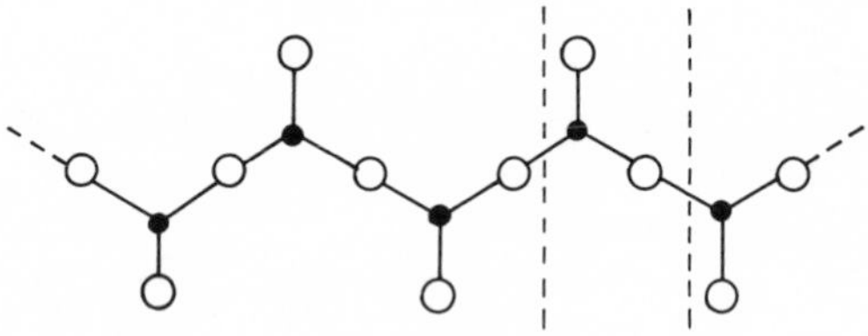

$(BO_2^-)_n$

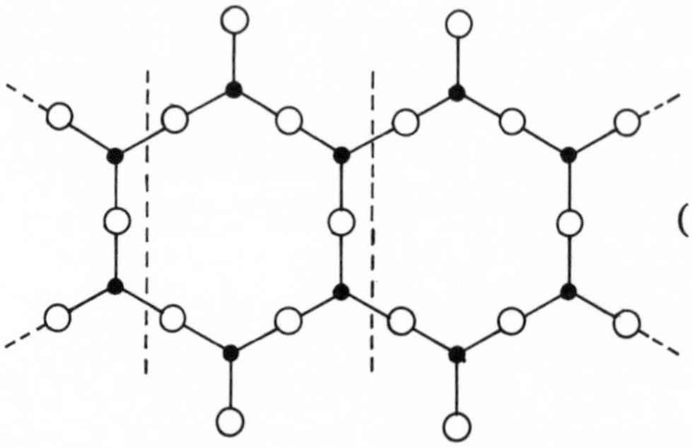

$(B_4O_7^{--})_n$

Fig. 9-19 Some possible types of polymers of a planar BO_3 group. Boron chemistry is complicated by the occurrence of tetrahedral BO_4 groups.

orbitals, and the number is normally equal to or less than that predicted on the ionic model.

2. When there is a choice between a layer lattice or "ionic" lattice, the layer lattice will be favored with more covalent compounds, particularly at low temperatures. If hydroxyl groups are present, hydrogen bonds tend to stabilize a layer lattice.

3. In a complex structure, formal oxidation numbers, i.e., the charges assigned to the atoms assuming an ionic model, tend to be satisfied by contributions and donations involving near neighbors.

4. When stoichiometry permits, highly charged cations will be as far apart as possible, and sharing of anions will not occur.

5. When sharing or polymerization does take place, cations tend to be as far apart as possible by forming linear bridges with the anions or by sharing only corners of their anion-coordinated groups. The effect is most important when the cation is small and assigned a large formal oxidation number.

Chapter 10

ISOMORPHISM AND SOLID SOLUTIONS

The term *isomorphism* tends to be used in two distinct ways. Literally, the term means "similarity of form." If any pair of substances is formed from atoms of similar dimensions and with similar types of chemical bonds, there is a very good chance that they will crystallize in crystals with identical structures and shapes. A typical pair is sodium nitrate ($NaNO_3$) and calcium carbonate ($CaCO_3$). In properties such as hardness, solubility, and fusibility, they bear no resemblance to each other, but the crystals are almost identical in interatomic dimensions and shape. It is not difficult to find the cause. The sodium and calcium ions are of similar size, $Na^+ = 0.95$ A and $Ca^{++} = 0.99$ A. The nitrate and carbonate ions are both trigonal-planar ions, the carbon and nitrogen atoms binding three oxygen atoms with an sp^2 hybrid. The C—O distance is 1.31 A and the N—O distance is 1.21 A. The volumes containing the formula weight or the molar volumes are 34.1 cm³ for $CaCO_3$ and 37.6 cm³ for $NaNO_3$. The main differences can thus be attributed to the stronger attraction of the anion and cation with double charges, Ca^{++} and CO_3^{--}, compared with Na^+ and NO_3^-.

The term *isomorphism* is commonly used in another sense. When a pair of cations or anions are of similar size, they are frequently found to replace each other in a crystal lattice with great facility as long as the condition of neutrality is satisfied. Such a process is better termed *substitution* or *solution* than isomorphous replacement. The replacement is readily visualized on the ionic model. If two cations are identical in size, and if the crystal is ionic, the affinity of either ion for a lattice position is identical. Thus in the reaction

$$X_{gas}^+ + Y^+Z_{crystal}^- \rightleftharpoons X^+Z_{crystal}^- + Y_{gas}^+$$

if the partial pressures of X^+ and Y^+ are the same, the chances of either ion's being in the crystal should be the same, and the resultant crystal should have the formula XYZ_2. Sometimes to show this substitution the formula is written $(XY)Z$, it being understood that the sum of $X + Y = Z$.

The above argument applies to (1) a pure ionic lattice, which does not exist; (2) ions of identical size, which also do not exist; (3) separation from the gas phase, which is of little practical importance. In most cases of interest, crystallization proceeds from a solution or from a melt or liquid, and further, the ions will differ in size by some amount and will also differ in electronegativity. The process may be visualized as follows:

$$X^+_{gas} + Y^+Z^-_{crystal} \xrightarrow{\text{II}} Y^+_{gas} + X^+Z^-_{crystal}$$

$$\text{I} \uparrow \qquad\qquad\qquad\qquad \downarrow \text{III}$$

$$X^+_{liquid} + Y^+Z^-_{crystal} \xrightarrow{\text{IV}} Y^+_{liquid} + X^+Z^-_{crystal}$$

Factors which influence the extent of the reaction IV involve: the solvation energy of X^+ in the liquid, step I; the relative lattice energies, step II; the solvation energy of Y^+ in the liquid, step III. The relative values of all these terms lead to the position of equilibrium of reaction IV and the extent of ionic substitution. These effects in real cases always lead to a preference or concentration of one or the other species.

A simple mineral pair exhibiting such behavior is Mg_2SiO_4 and Fe_2SiO_4, forsterite and fayalite. Both have similar crystal structures built from SiO_4^{4-} anions and Mg^{++} and Fe^{++} cations, whose sizes are 0.65 A and 0.75 A, respectively. Most natural olivines can be written with the formula $(Mg_xFe_y)SiO_4$ where $x + y = 2$, and we may also write the formula $(MgFe)_2SiO_4$. It is possible to prepare compounds with all possible ratios of x and y. In this case, when crystallization proceeds from a silicate melt, the crystals always show x/y greater than the value of x/y in the liquid. These data can be expressed in the melting or phase diagram below (Fig. 10-1). We may indicate the use of this diagram as follows. If a liquid of composition x is cooled, crystals commence to form at T_a, but the first crystals have a composition represented by point b on the composition coordinate. This pair of solids, Fe_2SiO_4 and Mg_2SiO_4, is said to exhibit a complete range of solid solutions, or the solids are completely miscible. It should be stressed that substances which are isomorphous in the sense of having similar form may or may not be miscible in the solid state.

In the substitution process, electroneutrality must be maintained. It is possible to obtain a substitution of the type

$$X^{++} + 2Y^{+}Z^{-} \leftrightharpoons X^{++}Z_{2}^{-} + 2Y^{+}$$

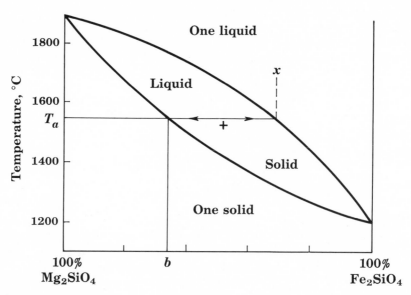

Fig. 10-1 Liquid-solid relations in the system Mg_2SiO_4–Fe_2SiO_4. For explanation, see text.

Consider the process as indicated in Fig. 10-2 below. In this case a vacancy or hole is created in the YZ lattice for each X^{++} ion entering. This tends to be a most unfavorable process, and such substitution is normally quite limited in extent.

If it is possible to exchange simultaneously two ions to maintain neutrality, ions with dissimilar charges may substitute with much greater facility. A good example is shown by the plagioclase feldspars albite ($NaAlSi_3O_8$) and anorthite ($CaAl_2Si_2O_8$). In this case the lattice can be considered to build from AlO_4^{5-} and SiO_4^{4-} anions which are polymerized in three dimensions. These groups are of similar size as are the calcium and sodium ions. Thus the substitution reaction

$$Na^+ + SiO_4^{4-} + CaAl_2Si_2O_{8_{solid}} \rightarrow Ca^{++} + AlO_4^{5-} + NaAlSi_3O_{8_{solid}}$$

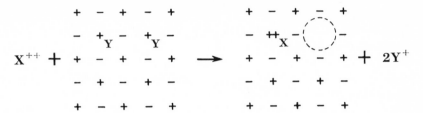

Fig. 10-2 Substitutions of cations of different charges causing a cation vacancy. Such solution is limited.

proceeds quite readily. Albite and anorthite, again, form a complete solid-solution series; and from a melt, the crystals formed tend to be enriched in $CaAl_2Si_2O_8$. The melting diagram (Fig. 10-3) is similar to Fig. 10-1.

The pair of feldspars $NaAlSi_3O_8$ and $CaAl_2Si_2O_8$ shows facile mixing; the pairs $NaAlSi_3O_8$ and $KAlSi_3O_8$ or $CaAl_2Si_2O_8$ and $KAlSi_3O_8$ show much less tendency. Again this is as expected on the ionic model. The potassium ion (1.33 A) is much larger than Na^+ or Ca^{++}, and extensive substitution should lead to distortion of the crystal lattice and a lowering of lattice energy. The thermal behavior of sodium-potassium feldspars is worth consideration and is indicated in Fig. 10-4.

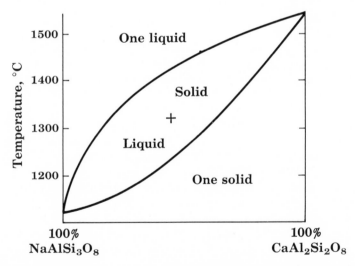

Fig. 10-3 Melting behavior in the system $NaAlSi_3O_8$–$CaAl_2Si_2O_8$. The solid solutions which form are the plagioclases. The diagram is similar to that of Fig. 10-1.

At high temperatures, where the lattice is expanded and atoms are vibrating with fairly large amplitudes, the pair $NaAlSi_3O_8$-$KAlSi_3O_8$ shows complete miscibility. As the temperature is lowered, the homogeneous solid solution unmixes into two mixtures nearer in composition to the pure compounds. A homogeneous phase x, formed at high temperatures, at temperature T_x forms two solid phases y and z, both nearer the pure end members. As the temperature is lowered further, the amount of solution becomes smaller.

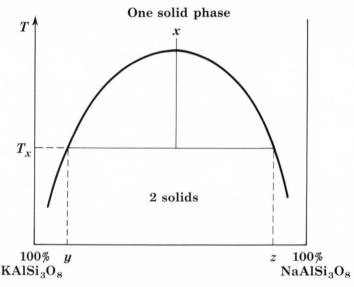

Fig. 10-4   The system $KAlSi_3O_8$–$NaAlSi_3O_8$, which demonstrates unmixing of a high-temperature homogeneous solid solution. For explanation see text.

Thus if two ions have rather different sizes, the amount of mixing is normally quite small, greater tolerance being exhibited at elevated temperatures. This phenomenon is a reflection of the greater entropy of a mixture compared to pure compounds (see page 7).

Elegant examples of extensive cation substitution are provided by zeolites. All zeolites are characterized by possessing a very open aluminosilicate framework built from polymerized SiO_4^{4-} and AlO_4^{5-} groups. The substances have very low densities compared with most silicates. In the aluminosilicate skeleton of zeolites there are rather large holes and channels, sometimes continuous, several angstroms wide. Neutral molecules can fill these channels (see page 128). The cations normally occupy sites close to the channels.

In the zeolite analcime ($NaAlSi_2O_6 \cdot H_2O$), the repeating structural unit contains twenty-four equivalent positions for the sodium ions, but at any given time only sixteen are occupied. If this material is placed in a solution of a salt other than sodium, some of the mobile sodium ions may be replaced. It is possible, for example, to produce the compound $Ca_{\frac{1}{2}}AlSi_2O_6 \cdot H_2O$ with only one-third of the sites occupied. This phenomenon is sometimes called *base exchange* or *ion exchange*, and materials which show this behavior with great facility are of considerable industrial and analytical importance, for it is often possible to selectively concentrate a given species.

The ionic model is often successful in predicting the relative tendency of foreign ions to enter a given host lattice. Some of these considerations which can be correlated with the lattice energy are:

1. Ions must be of approximately the same size as the substituted ion of the host lattice.

2. If two ions of similar charge are competing for a lattice position, the smaller is preferred.

3. If two ions of similar size are competing, that with the greater charge is preferred.

These rules introduced by V. M. Goldschmidt have been successful in explaining element fractionation in many cases of mineral formation, perhaps more successful than they should be. There are many exceptions, and one need not look far for the reasons. As mentioned above, if the ions were in the gas phase, we would expect the lattice-energy approach to be reasonable. But in most chemical and geological environments the competition involves some liquid phase. Consider a simple case of exception to these rules. If a zeolite or organic cation exchange polymer is placed in a solution of Li^+, Na^+, and K^+ chlorides, it is found that the order of affinity for the lattice is

$$K^+ > Na^+ > Li^+$$

the exact inverse of that predicted by the rules above. Electrochemists have shown that in aqueous solution where the ions are hydrated by ion-dipole forces, the order of effective size, i.e., the size of the ion plus its shell of water, is

$$Li^+ > Na^+ > K^+$$

To put the argument in another way, the small Li^+ ion interacts more strongly with the water molecules, and so to free the ion from the solvent requires more energy for Li^+ than Na^+ than K^+. Thus the ion with the smallest size (Li^+), while being most strongly bound in the solid, is also most strongly bound in the liquid through ion-dipole forces. These competing influences, almost always present in such a process, make any detailed calculations or predictions difficult, if not impossible, with our present knowledge. The order of magnitude of these effects is clearly seen from the figures in Table 10-1.

Table 10-1

Ion	Heat of solvation of ion in water, kcal	Lattice energy, kcal	
		Iodides	Fluorides
Li^+	121	174	240
	Difference, Li − Na, 26	Diff. 11	Diff. 27
Na^+	95	163	213
	Difference, Na − K, 19	Diff. 13	Diff. 23
K^+	76	150	190

Thus for the process

$$Li^+_{aq} + NaF_{solid} \rightarrow LiF_{solid} + Na^+_{aq}$$

$$\text{Heat of reaction} = -1 \text{ kcal}$$

but for

$$Li^+_{aq} + NaI_{solid} \rightarrow LiI_{solid} + Na^+_{aq}$$

$$\text{Heat of reaction} = +15 \text{ kcal}$$

The first process will be much preferred to the second.

One additional common type of exception should be noted. The zinc and magnesium ions are quite similar in size and identical in charge. In ionic compounds, i.e., compounds where x_A-x_B is large, one would expect considerable miscibility of zinc and magnesium compounds. But zinc is considerably more electronegative than magnesium, so that when these atoms combine with anions becoming progressively electropositive, covalent bonding will predominate with zinc compounds and determine the structure, while ionic bonding may still predominate with the magnesium compound. In zinc sulfide a tetrahedral sp^3 configuration is found, while in MgS the sodium chloride structure is found as anticipated from radius ratio considerations. In these compounds limited miscibility is observed due to the different configurations. Thus if two ions are of similar size and charge but differ appreciably in electronegativity, miscibility may be rather limited.

In a great number of cases the Goldschmidt rules work quite successfully, but as they are based on an oversimplified model, exceptions must be anticipated. To explain any given case, information must be available on every step of the substitution process— lattice energy, solvation energy, etc.—not one step, as implied by these rules.

Chapter 11

POLYMORPHISM

The phenomenon of polymorphism occurs whenever a given chemical compound exists in more than one structural form or atomic arrangement. The different structural types are polymorphs. The two forms of solid carbon, graphite and diamond, are typical.

The study of polymorphism has become of increasing interest to all concerned with the solid state. In the geological sciences we are particularly interested in the occurrence of such modifications, for there is a general tendency to find only one modification present in a given rock. This suggests that if we understood the conditions, physical and chemical, controlling the changes, we could gain important information on environments and processes in the earth's crust. The same is true for the modifications found in meteorites.

It is of great importance to appreciate that at any randomly selected pressure and temperature, the two forms will have different free energies, and, hence, one will be more stable than the other. Thus, given infinite time, we should find only one modification in our mineralogy museums. The interrelations of these modifications and the significance of rates of reactions and fundamental stability are well illustrated with reference to the forms of carbon and silica (SiO_2).

If the rate of graphite transformation to diamond and vice versa were rapid, and if we were to place a sample of pure carbon in a container so that the carbon could be subjected to any chosen pressure and temperature, we would find that diamond or graphite would be formed according to the information given in Fig. 11-1. Under all conditions, diamond would form below the line AB, and above the line graphite would form. Diamond is more dense than graphite, and when graphite is converted to diamond, the volume

140

of the carbon decreases by 36%. It will also be noted that as the temperature increases, the pressure necessary to form diamond also increases. Each form has a region or field of stability and a region where it is transformed or *tends* to undergo transformation in a region of metastability (see pages 6–9). There is also the unique line AB where the two forms are equally stable or are in equilibrium.

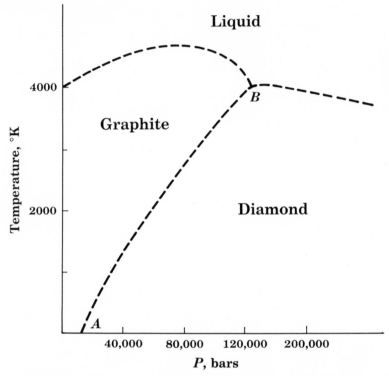

Fig. 11-1 Regions of stability of the various forms of carbon.

It is well known that under ordinary pressures and temperatures, diamond is a very stable substance and that its conversion rate to graphite is not capable of detection. Most polymorphic transitions show similar regions where rates of transition are sluggish, and thus forms are preserved in regions of metastability. The lack of rapid interconversions is a reflection of the large amount of energy required to break and rearrange the chemical bonds in the conversion reaction. Thus if diamond is to be converted to graphite, at least one chemical bond must be broken, all bond angles changed, and the state of carbon hybridization changed from tetrahedral sp^3 to trigonal-planar sp^2. The situation may be pictured as in Fig. 11-2a. An intermediate state of atomic rearrangement, the transition state, must occur and may have a much higher free energy than either stable or metastable state. Before reaction can proceed, an activation energy must be supplied by thermal energy. Thus, these processes become more rapid as the temperature increases, temperature above 1000°C in the case of diamond and

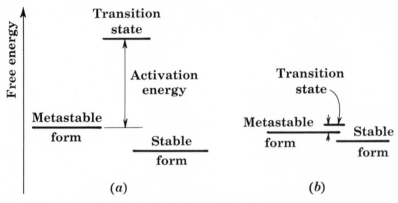

Fig. 11-2 Energy relations during polymorphic transformations: (*a*) for reconstructive transformations; (*b*) for displacive transformations.

graphite. In general, as the atomic rearrangement becomes greater, the rate of transition becomes slower, but there are numerous exceptions to this statement. A change of the graphite → diamond type, where physical properties and atomic arrangement of either form are quite different, is sometimes termed a *reconstructive transition*. The existing crystal structure must be taken apart and then rebuilt. A simple case of a transition state is indicated in Fig. 11-3.

A more complex array of polymorphs is shown by the modifications of SiO_2. The regions of stability are shown in Fig. 11-4. In this system there are four major modifications involving sluggish reconstructive transitions. While tridymite and cristobalite should form as quartz is heated, it is in fact possible to bypass these transitions and melt quartz directly if the heating is rapid. Coesite, a form of high density, requires greater pressures for formation at low temperatures than diamond. Although first produced artificially, it has now been found in large meteorite craters and included in diamonds. Most recently, a new form, stishovite, has been made

D + H–H ⟶ D···H···H ⟶ D–H + H

Fig. 11-3 Atomic arrangements during the reaction deuterium + H_2 → HD + hydrogen. In the transition state, a triatomic molecule with long, stretched bonds is formed. This transition-state molecule has higher energy than either initial or final state.

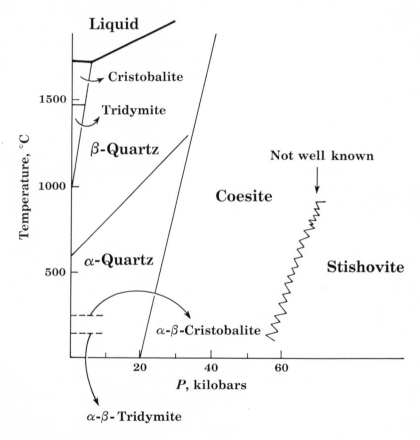

Fig. 11-4 Regions of stability of the various forms of SiO_2. The α-β transitions are rapid, but all other types tend to be sluggish.

with the rutile structure (Fig. 4-9) containing silicon in 6 coordination with oxygen.

Each of the forms quartz, tridymite, and cristobalite exists in two modifications, α and β. These transitions are rapid in both directions. The transition temperatures of α-β modifications of cristobalite and tridymite lie in the field of quartz and can be observed only on account of the sluggish nature of the major transitions at these temperatures. These α-β transitions involve rather minor structural alterations, slight changes in bond angles, lengths, etc., and the energy relations may be considered to look like those in Fig. 11-2b.

The rapid transitions involving slight structural changes are sometimes classified as displacive transitions. In the reconstructive transformation, new crystals form. In the displacive transformation, the old crystal form may remain with little outward reflection of the slight internal change. We shall see in Chap. 14 that this has important implications in the rate of reaction. Typical examples of such transitions are listed in Table 11-1.

Data on the conditions of formation of these and other polymorphs indicate some general trends:

1. As pressure increases, structures with high densities and large coordination numbers are favored; e.g.,

$$\text{Calcite} \rightarrow \text{aragonite}$$

$$\text{Sillimanite} \rightarrow \text{kyanite}$$

2. As temperature increases, structures with low densities and small coordination numbers are favored; e.g.,

$$\text{Quartz} \rightarrow \text{tridymite}$$

$$\text{Kyanite} \rightarrow \text{sillimanite}$$

3. The high temperature modifications frequently show the greatest symmetry; e.g.,

$$\text{Quartz (trigonal)} \rightarrow \text{cristobalite (cubic)}$$

Sometimes a structural modification may be prepared but is always in a region where another form is more stable. Such a modification is said to be a monotropic form. Boehmite is probably such a monotropic modification of $Al_2O_3 \cdot H_2O$, and maghemite a monotropic form of Fe_2O_3. Monotropic forms tend to be uncommon in geological environments.

It is obvious that polymorphs are valuable indicators of environments in which they form, but any simple interpretation must be approached with caution. It is well known that unstable forms

Table 11-1 Some Examples of Polymorphism

	Mineral	Density	Symmetry	Coordination nos.
$CaCO_3$	Calcite	2.715	Trigonal	C^3, Ca^6
	Aragonite	2.94	Orthorhombic	C^3, Ca^9
SiO_2	Quartz	2.654	Trigonal	Si^4
	Cristobalite	2.35	Cubic	Si^4
	Tridymite	2.27	Hexagonal	Si^4
	Coesite	3.01	Monoclinic	Si^4
	Stishovite	4.28	Tetragonal	Si^6
Al_2SiO_5	Kyanite	3.6	Triclinic	$Al^6Al^6Si^4$
	Sillimanite	3.25	Orthorhombic	$Al^6Al^4Si^4$
	Andalusite	3.15	Orthorhombic	$Al^6Al^5Si^4$
$Al_2O_3 \cdot H_2O$	Diaspore	3.4	Orthorhombic	Al^6H^2
	Boehmite	3.02	Orthorhombic	Al^6H^2
ZnS	Sphalerite	4.09	Cubic	Zn^4
	Wurtzite	4.0	Hexagonal	Zn^4
HgS	Cinnabar	8.18	Trigonal	Hg^6
	Metacinnabar	7.60	Cubic	Hg^4
C	Graphite	2.25	Hexagonal	C^3
	Diamond	3.51	Cubic	C^4

persist in metastable regions, but they may also be *formed* in a metastable region.

In simple cases, polymorphic changes would be expected and can be explained with reference to lattice energies. Consider the example of KCl. As mentioned in Chap. 4, the radius ratio of K^+/Cl^- is 0.734, a value almost identical with the ideal crossover value for $6 \rightarrow 8$ coordination. At 1 atm, potassium chloride crystallizes in the normal simple cubic sodium chloride structure. If a change to a body-centered CsCl structure is anticipated with increasing pressure, this latter form must have a smaller volume or larger density.

We may estimate the volumes. In the simple cubic structure a basic unit is as shown below. If R is the K—Cl distance $(3.149 \times$

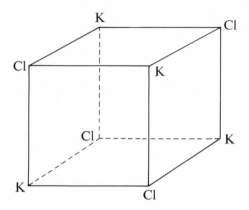

10^{-8} cm), the volume of this unit is

$$(3.149 \times 10^{-8})^3$$

Now every atom in this unit is shared with eight other identical cubes so that the contribution to the molecular volume is 1 atom. For a gram molecule of K^+ and Cl^- ions

$$V = 2 \times (3.149 \times 10^{-8})^3 \times \text{Avogadro's number}$$

$$= 2 \times (3.149 \times 10^{-8})^3 \times 6.025 \times 10^{23} \text{ cm}^3$$

$$= 37.6 \text{ cm}^3$$

In the body-centered modification the basic unit contributes two

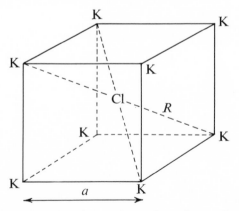

atoms or one KCl molecule. The cube edge a is related to the interionic distance $R = 3.149 \times 10^{-8}$ by

$$a = \frac{2R}{\sqrt{3}} = 3.635 \times 10^{-8} \text{ cm}$$

hence
$$V = (3.635 \times 10^{-8})^3 \times 6.025 \times 10^{23}$$
$$= 28.9 \text{ cm}^3$$

Thus the body-centered form is more economical with regard to space.

The lattice energy of either form is given by the relation

$$U = -\frac{Ae^2N}{R}\left(1 - \frac{1}{n}\right)$$

and, as mentioned previously, A is a summation constant depending only on the geometry of the lattice. A for the NaCl structure = 1.7626 and A for the CsCl structure = 1.74456. If we assume that R and n are the same for each lattice,

$$\frac{U_{\text{NaCl form}}}{U_{\text{CsCl form}}} = \frac{1.7626}{1.74456} = 1.01$$

The lattice energy of the stable 6-coordinated form slightly exceeds that of the 8, but as the volume of the latter is smaller, a transition must occur with increasing pressure. It can be shown that this pressure of transition P is given approximately by the relation

$$U_{\text{NaCl form}} - U_{\text{CsCl form}} = P_{\text{transition}} (V_{\text{NaCl form}} - V_{\text{CsCl form}})$$

Another important type of phase change is most common in complex crystals. We may illustrate this type with respect to the feldspar albite ($NaAlSi_3O_8$). This silicate contains a continuous network of SiO_4^{4-} and AlO_4^{5-} groups polymerized in three dimensions. Imagine that we can look along any given line of —Al—O—Si—O— atoms in the structure. In some specimens of albite we would find

—Al—O—Si—O—Si—O—Si—O—Al—O—Si—O—Si—O—Si—

but in another we might find

—Al—O—Al—O—Si—O—Si—O—Si—O—Si—O—Si—O—Al—O—

One species has a completely ordered arrangement of Al—Si atoms, while the other can be described as random. There are thus two distinct states, one fully ordered and one statistically random (for a simple example see Fig. 11-5). Between these distinct states, there are a very large number of states of varying degrees of order and disorder, each of which could be considered a separate polymorph.

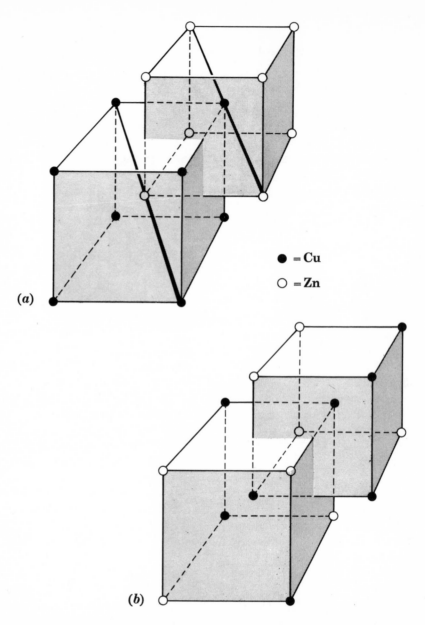

$= \text{Cu}$

$= \text{Zn}$

(a)

(b)

Fig. 11-5 A simple case of an order-disorder transformation in the alloy β-brass (CuZn). At low temperatures (a) all copper atoms and zinc atoms occupy unique positions at the corners of two interpenetrating cubes. At high temperatures (b) the atoms are randomly distributed on all lattice sites. The structure is body-centered cubic.

148

Such changes are often called order-disorder or second-order transitions. Unlike the first-order type discussed above, they do not occur at a sharp point, but disorder and entropy increase over a temperature range.

The rates of such transitions are highly variable, being quite slow in the case of albite except at temperatures approaching the melting point. In all cases the ordered structure is most stable at low temperatures, but some small amount of disorder is possible at all temperatures above absolute zero. At high temperatures, if melt-

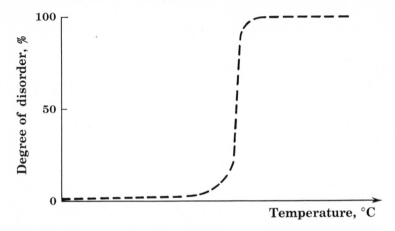

Fig. 11-6 Variation in the degree of disorder with temperature. Disorder starts above absolute zero, but there is normally a restricted temperature interval where it becomes virtually complete.

ing does not interrupt the process, the disordered modification becomes stable. A plot of the amount of order as a function of temperature might be as shown in Fig. 11-6.

In the case of the feldspars where silicon and aluminum atoms or oxide groups must move, the ordering process is a most difficult one to achieve. Thus metastable arrangements are common, and whenever the phases are grown rapidly at low temperatures, the high-temperature disordered state tends to be formed.

Finally let us note that through the earth pressures vary from 1 to 3 million atm. In the upper portions, pressure increases by about 300 atm for every kilometer increase in depth. Pressures of this order of magnitude can induce structural changes which at the surface are quite unfavored energetically, and there is no doubt that polymorphs will exist in the mantle of the earth in forms quite unfamiliar to us.   Slowly the experimental range is increasing, and today there is probably no modification occurring on the surface of the earth which has not or could not be synthesized in the laboratory.   At extreme pressures, even the electronic structures of the atoms may be changed.   For example, metallic cesium, one of the alkali metals, undergoes a very marked volume reduction at high pressures.   The electronic structure of the normal atom is

$$1s^2\ 2s^2\ 2p^6\ 3s^2\ 3p^6\ 3d^{10}\ 4s^2\ 4p^6\ 4d^{10}\ 4f^0\ 5s^2\ 5p^6\ 5d^0\ 5f^0\ 6s^1$$

It will be noticed that empty levels in the fourth and fifth quantum group are available, and transitions involving the placement of fifth- and sixth-group electrons in inner levels are possible.

Chapter 12

DEFECTS IN CRYSTALS
AND THE COLOR OF CRYSTALS

Defects in Crystals

There is perhaps more fascination in what things do than in what they are, and while the static geometry of crystals is of great significance, when many properties and reactions of crystals are considered, very minor imperfections or defects become of major importance. Frequently, the slightly displaced atom, ion, or electron will react with much greater velocity than the same species in a regular lattice site. This chapter must be considered as only a very slight introduction to a subject which has received much attention from all in the broad field of solid-state physics and chemistry.

The model of the ionic or covalent crystal that we have discussed earlier, involving an orderly array of atoms, fails to account for some properties and features of the reactivity of solids. There is always present a small number of species which are more mobile than anticipated. The "perfect" model we have discussed earlier is most successful at the absolute zero of temperature. At all higher temperatures defects are present.

Two types of defect are always present in ionic crystals, the so-called Frenkel and Schottky defects. These types are shown in Fig. 12-1. When a Frenkel defect forms, an ion leaves the normal lattice site and occupies some "interstitial" position. A Schottky defect is formed when an ion leaves its lattice site and migrates to the surface or, more simply, when a lattice site is vacant. Both defects thus leave "holes" in the orderly array of the crystal. The number N of these defects increases exponentially as the temperature of the crystal increases and obeys a law of the form $N \propto e^{-E/RT}$, where T is the absolute temperature, R the gas constant, and E

the energy needed to create the particular type of defect, the activation energy for defect formation. We may note that the formation of defects is quite similar to the order-disorder phenomena discussed in Chap. 11, and the defect-forming process could be considered as a second-order phase transition. But while the disordering process frequently becomes complete before a crystal melts, the introduction of a large number of defects weakens a crystal, and the number never becomes large because of interruption by melting.

In any given crystal, while both types of defect may be present, one will be easier to form (that with the smaller E value) and will predominate. Further it should be noted that Frenkel defects in an ionic lattice must form in pairs to maintain electric neutrality, and thus will be favored in crystals where cation and anion are of similar size, and hence are bound more or less equally in the lattice. Such is the case in KCl. When the ions are dissimilar in size—for example, NaCl—Schottky defects will form much more easily, and the small ion will be favored in the interstitial site. In the sodium chloride structure an interstitial sodium ion will be situated as in Fig. 12-2, and it should be recognized that in this position there is room for this smaller species, although energetically it is not a preferred position.

Ionic crystals show some capacity for conducting electricity, and it is a simple matter to demonstrate that ions may be mobile. If we grow a layer of radioactive KCl on one end of a large KCl crystal and heat the crystal for a prolonged period, we shall find that the activity is evenly distributed throughout the crystal. The diffusion processes and conductivity of electricity also vary exponentially with temperature, and these measurements can actually be used to measure the number and nature of defects present. Some

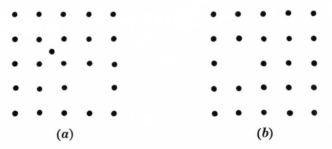

(a)   (b)

Fig. 12-1 (a) A Frenkel defect where an atom leaves its lattice site and migrates to an interstitial position; (b) a Schottky defect or vacant lattice site.

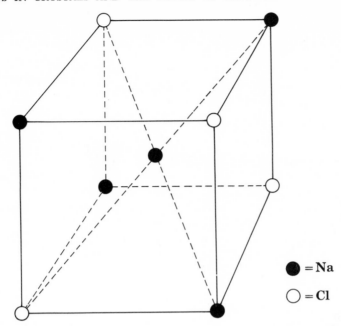

Fig. 12-2 Position of an interstitial sodium ion in sodium chloride.

striking experiments have been performed to demonstrate the mobility of species in solids. In Table 12-1 is listed the concentration of interstitial ions in AgBr as a function of temperature.

Table 12-1 Concentration (%) of Silver Ions in Interstitial Positions in Silver Bromide (Melting Point, 426°C)

T, °C	Concentration, %
426	2
300	0.4
250	0.18
210	7.6×10^{-2}
20	8.3×10^{-4}
−180	10^{-20}

If two slabs of silver chloride (AgCl) are placed between two slabs of silver and these are connected to a battery as shown in Fig. 12-3, it is found that, as current passes, slab D of silver grows at the expense of silver slab A, while the original AgCl layers remain

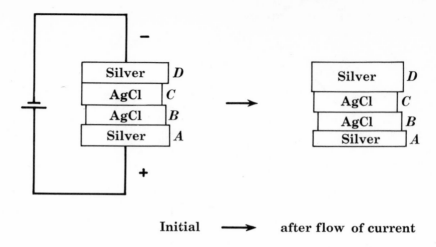

Initial   ⟶   after flow of current

Fig. 12-3 Experimental setup to demonstrate the mobility of silver ions in silver chloride.

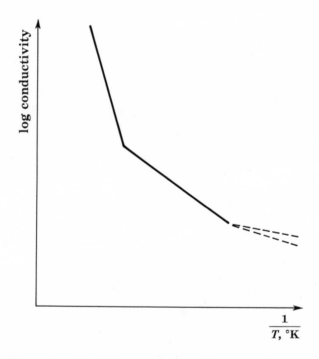

Fig. 12-4 Form of conductivity-temperature relation shown by lead iodide. For explanation see text.

unchanged.   Silver ions are diffusing through the crystal, while the chloride ions are not carrying any current and are immobile.

If the number of defects is proportional to $e^{-E/RT}$ and if electric conductivity is also proportional to the number of defects, we expect a relation of the form

$$\log \text{conductivity} \propto \frac{1}{T}$$

This indeed is found to be the case.   In Fig. 12-4 is indicated the type of conductivity-temperature behavior found with lead iodide (PbI_2).   The conductivity curve has three distinct linear portions. That part at lowest temperature has the smallest activation energy— the conducting species move easily.   Further, this portion is not constant from crystal to crystal but depends on preparation or history.   It is said to be "structure-sensitive" and may represent conductivity along cracks, conductivity due to impurities, etc.   The next two steeper sections indicate a more difficult ionic diffusion process, and these are found in all crystals of lead iodide.   The first step is due to conduction by Pb^{++} ions and the steepest step at highest temperatures is due to conduction by both Pb^{++} and I^- ions.

It is quite simple to see why the defect ions are more mobile than the same ions in the regular lattice position.   When the defect moves, the main barrier involves squeezing past ions to reach the next interstitial site.   When the regular lattice ion moves, it must be taken out of its lattice site and then squeezed between its neighbors. The situation is indicated in Fig. 12-5.

It is possible experimentally to distinguish conductivity in a solid due to positive ions, negative ions, or electrons.   The ionic carriers will be defect species, but as we have mentioned already (page 68), substances like graphite, where delocalized π electrons are present, are electronic conductors.   In a crystal with ionic or σ covalent bonds, the usual treatments assume the electrons are either concentrated on ions or atoms or are localized in bonds between two atoms.   This is never exactly true.   For example, in diamond the bonds are formed from sp^3 hybrid orbitals on each carbon atom. If we consider any three carbon atoms in a row, almost all the exchange interaction or covalency is between atoms 1 and 2 and 2 and 3.   But there is a very small contribution due to overlap of the sp^3 orbital on atom 1 with that on atom 3.   In other words a small quantity of the electron cloud is spread through the crystal and may contribute to electronic conductivity.   In the case of most normal covalent and ionic crystals, this effect makes a much smaller

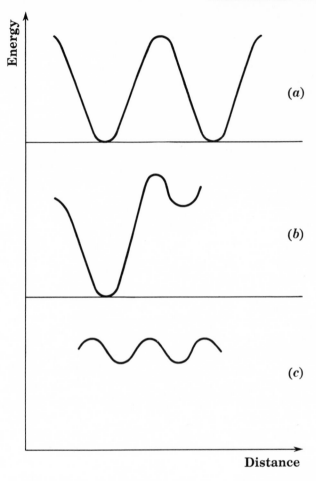

Fig. 12-5 Energy changes in moving ions in a lattice: (*a*) one equilibrium position to another; (*b*) an equilibrium position to an interstitial position; (*c*) one interstitial site to another.

contribution to the conductivity than do defect contributions. The delocalization electronic conductivity of crystals increases with increasing temperature.

Many metal sulfides are quite metallic in appearance, particularly sulfides of the transition metals. Again many are electronic conductors showing an analogy with metals. In many transition-metal compounds delocalized π electrons are to be expected through π bonds formed with the *d* orbitals of the metal. A simple example of such a situation is indicated in Fig. 12-6, where metal d orbitals form a continuous π chain with p orbitals of a nonmetal. Such bonds lead to conductivity as in graphite. The possible types of such $d\pi$ interactions are numerous and complex.

In the examples mentioned above the defects occur in pure crystals of ideal composition. In such crystals two other types of imperfection are common: mosaic defects, or defects caused by dislocations, and stacking defects. In recent years the importance of dislocations in consideration of mechanical properties and crystal growth has been realized. We shall return to these in Chap. 13 in dealing with the formation of crystals.

When impure crystals are considered, the possible types of defect are numerous. One of the most striking types, and one most common in minerals, is illustrated by the colored alkali halides. If potassium chloride is heated in potassium vapor, the crystal becomes blue. Careful chemical analysis indicates that these crystals have a formula

$$K_x Cl \qquad x > 1$$

Numerous studies of electric and optical properties indicate that the solution of the excess metal produces mobile electrons in the crystal and that reaction proceeds according to the equation

$$K_{vapor} \rightarrow K^+_{crystal} + e_{crystal}$$

If a blue crystal is placed between an anode and cathode at a moderate temperature, the blue color will be seen to migrate to the anode,

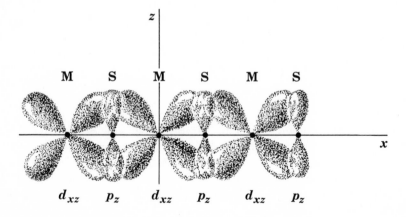

Fig. 12-6 A molecular $d\pi$ orbital (cf. graphite) formed along a row of transition-metal and sulfide ions by interaction of d_{xz} and p_z orbitals.

perhaps one of the most striking of all demonstrations that some species in crystals are capable of rapid motion. The most common explanation of the situation of the added potassium ion and electron is that the potassium ion occupies a normal lattice site or interstitial site, while the electron occupies an anion vacancy and thus acts as a small anion. The electron in this position is known as an F center or color center. It is easily moved either optically or electrically. If the concentration of these semifree electrons in a crystal becomes sufficiently large, in place of the single electron in an anion vacancy, a pair of electrons may be trapped producing an F' center.

F and F' centers with their accompanying coloration can be produced in natural minerals in other ways. For example, if a KCl crystal is irradiated with energetic particles (x-rays or α particles), the blue coloration forms rapidly. In this case the radiation has caused reduction via the process

$$Cl^- + \text{energy} \rightarrow Cl + e$$

The electron thus set free may then be trapped in an anion vacancy. If a crystal with F centers, but still of composition KCl, is heated, the electrons rapidly find their way back to a neutral chlorine, and the color is thermally bleached. Many natural minerals such as fluorite (CaF_2), calcite ($CaCO_3$), and halite ($NaCl$) show such coloration produced by long exposure to radiation in a natural environment.

Most students exposed to elementary chemistry have observed that when zinc oxide is heated, it becomes yellow, an observation used in testing for zinc compounds. The coloration is again due to production of F centers. When ZnO is heated, some oxygen is evaporated from the crystal, leaving an excess of metal in the crystal and electrons in F centers. When the solid is cooled, the oxygen is reabsorbed, and bleaching occurs.

In many substances, an excess of nonmetal can be formed, and perhaps one of the best examples is FeO. In this case it may be impossible to produce the stoichiometric compound. In such compounds cation vacancies occur. It must be considered that some oxygen anions do not have a full eight-electron shell in such a crystal. The region from which the electron is missing, or the hole in the oxygen valence shell, is described as a "positive hole," a potential electron trap. Conduction can now occur by movement of the "positive hole."

Titanium monoxide (TiO) can have compositions ranging from $TiO_{0.6}$ (excess metal) $\rightarrow TiO_{1.35}$ (excess oxygen). $TiO_{0.6}$ will contain

F centers and be an electron or n conductor, while $TiO_{1.35}$ will contain positive holes and be a p conductor. It is worth noting that substances with positive-hole defects are commonly those containing transition metals of variable valence.

The formation of some classes of solid solutions (mixed crystals) is often associated with increased defect formation and ionic conductivity. Silver chloride (AgCl) which crystallizes in the sodium

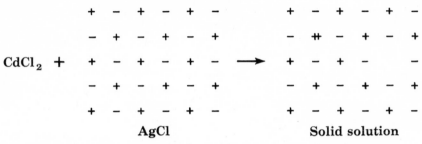

AgCl   **Solid solution**

Fig. 12-7   Cation vacancies formed in AgCl by solution of $CdCl_2$.

chloride structure will dissolve cadmium chloride ($CdCl_2$) without any change in structure. The situation is illustrated in Fig. 12-7. It will be noticed that for every cadmium ion placed in the lattice, a cation vacancy is created. In nickel oxide (NiO) if some of the doubly charged nickel cations are replaced by lithium (Li^+) ions an enormous increase in conductivity is observed. For example, a crystal with 10% of the nickel ions replaced has one hundred million times the conductivity of pure nickel oxide.

Similar impurity effects can be produced in covalent crystals. As we have seen, silicon forms a diamond-type lattice (Fig. 5-2), each silicon being linked to four other silicon atoms by sp^3 tetrahedral hybrids. If we replace some of the silicon atoms by phosphorus, we are essentially adding to the crystal extra electrons according to the electronic structures below:

$$Si \qquad 1s^2\, 2s^2\, 2p^6\, 3s^2\, 3p^2$$
$$P \qquad 1s^2\, 2s^2\, 2p^6\, 3s^2\, 3p^3$$

These extra electrons are accommodated in higher energy molecular orbitals and are more mobile than the valence electrons. The phosphorus-substituted crystal thus becomes an electronic (n) conductor. If, however, an aluminum atom is added with structure

$$Al \qquad 1s^2\, 2s^2\, 2p^6\, 3s^2\, 3p^1$$

a mobile positive hole in the valence electrons is created and this crystal is a p conductor.

Another very common situation is responsible for enhanced electron mobility in many minerals. Magnetite (Fe_3O_4) contains both ferric (Fe^{3+}) ions and ferrous (Fe^{++}) ions. Were there no abnormalities (see section on color), this substance should be pale green. In fact, magnetite is black and metallic in appearance. The electrons responsible for these physical anomalies are produced by interaction of the metal ions in different valence states. If these are sufficiently close, an electron can be readily exchanged:

$$Fe^{3+} + e \leftrightarrows Fe^{++}$$

and this exchange causes the anomalous conductivity and color. The same phenomenon causes the intense black coloration of biotite, one of the mica structure silicates. In the mica structure (Fig. 9-16) the iron atoms within the sheets are much closer than atoms in separate sheets, so the electron exchange is concentrated within the sheets, and the electric conductivity and optical absorption are quite different in different directions.

The Color of Solids

A casual glance at any mineral collection indicates that a great number of minerals are colored. Further, we could very rapidly separate uncolored (white) minerals from those colored on the basis of chemical composition. The great majority of colored inorganic materials would contain transition metals.

The visible spectrum (that detected by the human eye) consists of the following wavelengths (in angstroms):

Violet	4000–4240
Blue	4240–4912
Green	4912–5750
Yellow	5750–5850
Orange	5850–6470
Red	6470–7000

At shorter wavelengths we have the ultraviolet region of the spectrum, and at longer wavelengths the infrared region. For a mineral to appear green in color, radiation in the violet and orange-red regions of the visible spectrum must be absorbed.

The most common cause of coloration in inorganic materials is the presence of a transition metal, and, further, it is well known that whenever an atom with a partially filled d or f electron shell is present, there is a high probability that its compounds will absorb in the visible region.   It is possible to explain why this occurs.

Let us consider one of the simplest transition metal ions, Ti^{3+}. The structure of this ion is

$$Ti^{3+} \qquad 1s^2\, 2s^2\, 2p^6\, 3s^2\, 3p^6\, 3d^1$$

As we have mentioned earlier, the $3d$ level will hold five pairs of electrons with opposed spins:

(⇅) (⇅) (⇅) (⇅) (⇅)

as in the zinc ion Zn^{++}.   In a free atom, these five d levels have equal energies, and thus an electron may move from one to another without absorption of energy.

In Fig. 2-5 the shape of the electron cloud associated with each of the d orbitals is illustrated.   Now let us place the Ti^{3+} ion in a crystal with the sodium chloride structure where it will be surrounded by an octahedral grouping of six anions.   A similar situation would arise if we placed the ion in water where it would be again surrounded by an octahedral group of six water molecules with the oxygen atoms directed toward the central ion.   The ion is thus placed in the electrostatic field of the anions, and this field can disturb the energy of the various d levels.

In Fig. 12-8, a Ti^{3+} ion in water is illustrated.   If we consider the xy plane of the grouping, we shall see that the d electron orbitals are distributed as in Fig. 12-9.   One of the d orbitals in this plane has its lobes directed along the x and y axes toward the oxygen atoms.   The other d orbital has its lobes directed between the oxygen atoms.   Clearly, if we place an electron in the d orbital directed toward the oxygen atoms, it will be repelled by the pairs on the oxygen more than if it were placed in the other d orbital.   Thus the "ligand" or "crystal" field has "split" the degenerate d level. The orientation of orbitals in the yz and xz planes is shown in Fig. 2-5.   In effect, two of the d orbitals are in unfavored positions relative to the other three.   The d level has thus been split into two sublevels:

(↑) ( ) ( ) ( ) $\xrightarrow{\text{field}}$ ( ) ( )   High energy

(↑) ( ) ( )   Low energy

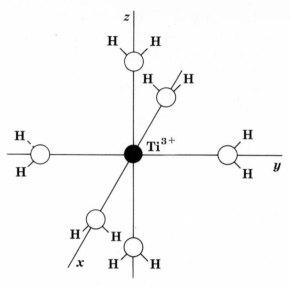

Fig. 12-8 Octahedral environment of Ti^{3+} in aqueous solution. Lone-pair electrons on oxygen atoms are directed along the xyz coordinates.

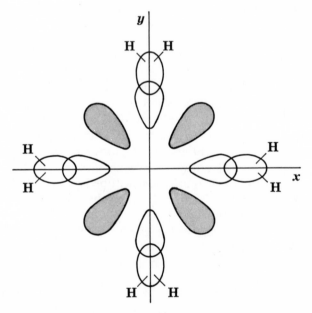

Fig. 12-9 The xy plane of a transition-metal ion in an octahedral environment in water. The $d_{x^2-y^2}$ orbital is unshaded, the d_{xy} orbital concentrated between the water molecules is shaded (Fig. 2-5).

162

The single electron of Ti^{3+} will now be located in one of the three low-level orbitals. The difference in energy of these upper and lower levels ·is commonly similar to the energy of a photon or radiation in the visible region of the spectrum, and this energy may be absorbed and cause the electron to jump from one level to the other.

With Ti^{3+} we would expect a single absorption band caused by this excitation. This is observed (Fig. 12-10) and causes the purple-red coloration so common with titanium compounds.

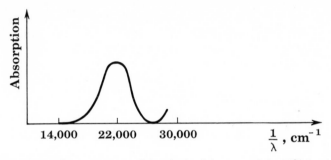

Fig. 12-10   Absorption spectrum of the single *d* electron of the Ti^{3+} ion in water.

Two questions in the above discussion require some comment. First in Fig. 12-10 it will have been observed that light is absorbed over a rather broad range of wavelengths. This means that a large range of energies must be involved in this rather simple process. The reason for this breadth is that the anions or water molecules surrounding the ion are vibrating due to thermal agitation. As an anion swings toward the ion, the difference in energy of the two types of *d* orbitals—the splitting—increases. As the anion swings farther away, the splitting decreases. The maximum in the absorption curve thus corresponds approximately to the splitting by the average position of the anions. If we could measure the spectrum at very low temperatures where the thermal vibrations are reduced, the absorption band would become much narrower.

The second question which may be raised is why we chose the orientation of orbitals selected in Fig. 12-9. In part, the answer to this question is that so far we have assumed an ionic model to explain the splitting. But there must always be a covalent contribution, and if covalent bonds form in an octahedral environment, the titanium ion will use a d^2sp^3 hybrid for these bonds formed from two of the *d* orbitals. These are the ones chosen to be axially directed and are the high-energy orbitals.

The position of the absorption band or bands shown by a transition-metal ion depends on the distance of the negative charges from the central cation and on the electronegativity of the surrounding groups. As groups come closer (e.g., during compression of a crystal) or as the bond from ligand to metal becomes more covalent, the splitting increases, and more energy is required to excite the electrons. Thus the absorption bands move to shorter wavelengths.

The complexity of the d electron spectrum changes with the number of electrons in the unfilled shell. The prediction of the complexity of the spectrum of an ion with many d electrons may be quite complex (see Orgel reference, page 193). As well, other "excited states" may add to the spectrum if changes such as

$$Ti^{2+}, 3d^2 \rightarrow Ti^{2+}, 3d^1\, 4s^1$$

have energies in the visible range.

Ions such as manganous and ferric show very little interaction with visible light. Both these ions have the same structure:

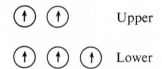

When this level is split, we then have

(↑) (↑)   Upper

(↑) (↑) (↑)   Lower

Now electrons can only occupy the same orbital if the spins are opposed, and excitation can only occur if the spin orientation changes. The transition is said to be "spin forbidden." Such ions do show spin-forbidden transitions, but the process is unfavorable as electrons are reluctant to change spin, and the absorption bands are of feeble intensity. Thus manganous salts show only a very pale color, and ferric salts such as the nitrate are colorless. If a pure ferric compound is highly colored, some electron delocalization phenomena, or simple photochemical reduction process, must occur (see pages 157 and 168).

In summary, the color of transition-metal compounds is caused by excitation of d electrons in electrostatic crystal fields which

remove the essential degeneracy of the five d levels.   The nature of
the spectrum, number of bands, and intensity of absorption depend
largely on the number of d electrons and the geometry of the coordinate group.   The position of the bands depends on the nature
of the species producing the negative field.   Thus color is not at all
characteristic of a given ion; only the complexity of the spectrum
is characteristic.   If, however, we compare colors of compounds
where the same atom surrounds the metal ions, the color tends to
be quite similar.   Thus the aqueous ferrous ion ($Fe^{++} \cdot 6H_2O$),
solid ferrous sulfate ($FeSO_4 \cdot 6H_2O$), olivine (($MgFe)_2SiO_4$), actinolite
($Mg_2(FeMg)_5S_{18}O_{22}(OH)_2$) all show a similar green color.
    The nature of the splitting of the d levels depends on the ge-
ometry of the group surrounding the ion; the spectrum in a tetrahedral field will differ from that in an octahedral field.   In a
tetrahedral field, because the transition-metal ion will use a d^3s
tetrahedral hybrid, the d level is split as indicated below:

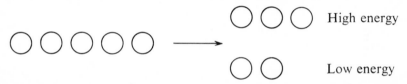

(Note that ions such as Ti^{4+} and Zn^{++} where the d shell is empty
and full, respectively, will show no color because of d-shell transitions.)
    In the examples discussed above, the electron configuration with
regard to spin is unchanged.   If the splitting becomes very large,
a new situation may arise as indicated below for a ferrous ion Fe^{++}.

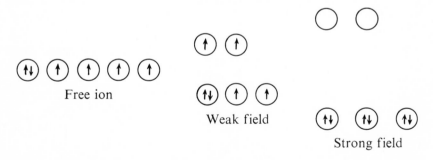

In the strong field, the difference in energy of the two parts of the
split d level is so great that the electron-electron repulsion caused
by putting both electrons in the same orbital is compensated by the

lower energy of low-energy orbitals.   When this occurs, the magnetic properties of the ion, a function of the number of unpaired electrons, are drastically changed.   Further, the ion shrinks in size by a considerable amount, for the anions can now move in without introducing a great deal of extra repulsion.   We may note that such coupling of electrons leads to a smaller ion and hence another cause of polymorphism at high pressures.

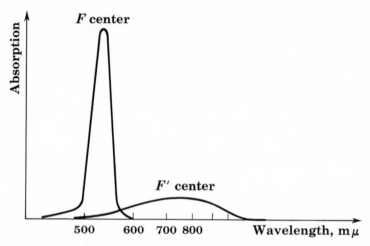

Fig. 12-11   Absorption bands caused by F and F' centers in KCl.   The wavelength scale is logarithmic.

If a grouping around a cation is not symmetrical or electrostatically symmetrical—e.g., a slightly distorted octahedron—the splitting is different in different directions in the crystal, and all five orbitals may differ in energy.   If such a crystal is viewed in polarized light, it will show different colors in different directions, or be pleochroic.   Pleochroism could not occur if the crystal field were symmetrical.

As already discussed (page 158), the color of some naturally occurring inorganic compounds of metals with complete electron shells may be associated with defects such as F and F' centers, and these same colorations can frequently be produced by x-radiation, etc.   The mineral fluorite (CaF_2) is most commonly blue in color, and this color is easily thermally bleached.   The bleaching is caused by thermal release of free electrons in F centers.   Illumination with wavelengths in the absorption band of the F- and F'-center crystal may also free the electrons from their traps and cause bleaching.

The absorption bands of F and F' centers in KCl are shown in Fig. 12-11.

Just as the rather mobile electrons in defects can be excited and absorb visible radiation, mobile electrons in π orbitals or delocalized electrons may absorb strongly in the visible region of the spectrum. The common organic dyestuffs typically contain a large number of π electrons. Absorption in graphite which covers the entire visible spectrum involves these mobile π electrons, while diamond with no such electrons is completely transparent to visible radiation. There is little doubt that coloration of many sulfides is also partly caused by such mobile electrons for sulfur (selenium and tellurium); all readily form π bonds with transition metals. In many transition-metal compounds, metal-metal π bonds may occur and cause the absorption in the visible region (see Fig. 12-6).

The question may well be raised in the case of π-electron excitation as to where the electrons are excited, and here we may introduce another quite general concept. We have mentioned previously that in a molecule such as H_2, the bonding electrons occupy a molecular state we can describe by combining the wave functions for the atomic states:

$$\psi_{\mathrm{mol}} = \psi_{1s} + \psi_{1s}$$

But whenever states are combined in this fashion, the number of possible states remains unchanged. The state we have just described with the positive combination sign is a bonding state. There is also another coupled solution:

$$\psi_{\mathrm{mol}}^* = \psi_{1s} - \psi_{1s}$$

known as an antibonding state. This state leads to no chemical combination. The electron density contours for electrons in each state are illustrated in Fig. 12-12. The antibonding state leads to no concentration of electrons between the nuclei. Now in a molecule such as graphite, the π states are described by a combination of C_{2p} orbitals through the entire crystal, and N-bonding π orbitals are formed where N is the number of carbon atoms. These orbitals are completely filled, but as well we must produce N-antibonding π^* orbitals, and excitation into these energy states is possible. It is only when a large collection of such states is present, continuous through the molecule or crystal, that visible excitation commonly occurs.

As we move into the short-wavelength region of the visible spectrum or into the ultraviolet, the more energetic photons can cause much more difficult processes to occur. Those causing absorption which impinges on the visible region of the spectrum are frequently due to what are called *charge transfer* or *photochemical oxidation reduction* processes. For example, the intense colors shown by many metallic oxides and sulfides are partially due to such changes. Thus in ferric oxide we may have

$$Fe^{3+} + O^{--} + \text{photon} \rightarrow Fe^{++} + O^{-}$$

All inorganic materials exhibit such spectra, but only those with easily reduced cations or oxidized anions will produce this absorption

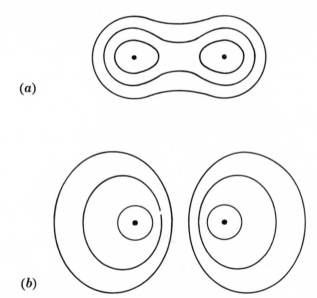

(*a*)

(*b*)

Fig. 12-12 Electron density contours in (*a*) the bonding state of the hydrogen molecule and (*b*) the antibonding state. In the latter, there is no concentration of electron density between the nuclei.

in the visible region. For example olivine (($MgFe)_2SiO_4$) is green in color. This color is caused by the Fe^{++} ion's absorbing the red end of the spectrum in a *d* electron excitation and by charge-transfer absorption occurring at the blue end of the spectrum. Again, the coloration of many ferrous-ferric compounds, e.g., magnetite and biotite, is caused by the process

$$Fe_1^{++} + Fe_2^{3+} + \text{photon} \rightarrow Fe_1^{3+} + Fe_2^{++}$$

We have thus outlined some of the causes of optical absorption in the visible region. Much is yet to be learned and in many cases a number of causes may be superimposed. For example in the mineral epidote we may have ferrous ion d absorption, ferrous-ferric exchange absorption, and charge transfer all causing the resultant green coloration. In any compound, the spectrum can tell us much about the elements present, their oxidation states, the types of defect present, and features of the chemical bonds.

The intensities or efficiency of the various optical absorption processes may be quite different. The transition-metal spectra (d electron excitation) tend to be many orders of magnitude weaker than charge-transfer spectra. The spin-forbidden transitions of the manganous and ferric ions are again powers of 10 weaker than the normal transition-metal bands.

Chapter 13

THE FORMATION OF CRYSTALS

As we have already mentioned in Chapter 1, crystallization may proceed by different paths:

1. Vapor → solid, condensation
2. Solution → solid, precipitation
3. Melt → solid, freezing
4. Solid A → solid B, transformation

All these processes have in common the condition that for crystallization to proceed, the final crystals must have lower free energy than the initial state of the system. This condition implies that for path 1 the vapor pressure of the system must exceed the vapor pressure of the solid crystal; for path 2, the concentration of the species in solution which form the solid must exceed the concentration in a solution of these species formed by dissolving the solid; for path 3, the temperature of the melt must be less than the melting point of the solid; for path 4, solid B must possess lower free energy than A and hence a smaller solubility in a given solvent and a lower vapor pressure.

The process of forming a crystalline phase involves two steps: first the formation of a new nucleus, and second the growth of this nucleus to form a particle of appreciable size.

Nucleation

It is quite remarkable that crystals form at all. Let us imagine that we are slowly evaporating a solution so that eventually it becomes supersaturated and a solid phase appears. When the solid forms, initially a small group of molecules or ions must come together to form a nucleus which will then grow. The very small initial nucleus must be extremely unstable, for, as we have discussed

170

earlier (page 53), it possesses a very large surface energy. Imagine, for example, a single unit of the sodium chloride structure (Fig. 13-1). It will be seen in this that only one-half of the normal valence forces are satisfied, and such a small crystal has enormous free energy and hence solubility. Such a small nucleus will tend to dissolve unless it can grow very rapidly to achieve a size where the surface energy becomes smaller in relation to the total energy of the crystal. When the newly formed nucleus reaches a size where it is equal in free energy to the initial material, it has reached what is known as the critical size and can now survive to grow. The energy relations of the various states of a crystallizing solution are schematically shown in Fig. 13-2. It is possible to show theoretically that the size of the critical nucleus is a function of the amount of supersaturation of the solu-

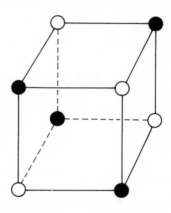

Fig. 13-1 A single unit of the NaCl structure which might form a highly unstable nucleus.

tion, and hence nuclei form much more easily as the degree of supersaturation increases. This effect is in turn reflected in the type of precipitates formed:

> If supersaturation is large → many nuclei, gel
>
> If supersaturation is moderate → fine crystals
>
> If supersaturation is small → few large crystals

Thus the best large crystals are produced if the degree of supersaturation is small or if a seed crystal is added to a very slightly supersaturated solution. The production of very large crystals, several feet long in some rocks, may be a reflection of a very low degree of supersaturation.

Not all nuclei form under conditions as outlined above, but frequently nucleation is induced on some surface, e.g., the walls of the vessel or dust particles, present in the system. The surface on which nucleation is induced will have, in the most favorable cases, some interatomic distance comparable with that in the solid being formed. A striking case of such induced nucleation is shown by the thermal decomposition of manganous nitrate. This compound decomposes into oxides of nitrogen and a manganese oxide. If the

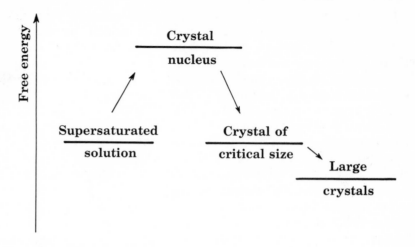

Fig. 13-2 Relations between the free energy of a saturated solution, crystal nucleus, small stable crystal, and large crystals.

nitrate is precipitated on TiO_2 (rutile) before decomposition, the solid product is MnO_2. But if the nitrate is precipitated on Al_2O_3 (corundum), the product is Mn_2O_3. The dimensions of MnO_2 and TiO_2, and Al_2O_3 and Mn_2O_3, are sufficiently similar that the path of the reaction is controlled by the ease of nucleation, not necessarily a path leading to the lowest free energy. In effect, the surface induction eliminates the necessity of formation of a nucleus of critical size.

When solid-solid transformations are occurring, the difficulty of nucleation often controls the rate of the process. In second-order phase transformations (see page 149), this barrier is not present, and such changes are often rapid and reversible as compared with first-order transformations. The importance of nucleation in first-order transitions is well demonstrated by the stages in the crystallization of amorphous silica at low temperatures in the presence of water. At temperatures below 870°C, the most stable solid form of silica is the polymorph quartz. But amorphous silica is somewhat similar to cristobalite in structure, and although this form is less stable than quartz, it tends to form initially and then, in turn, pass to the stable phase. This common phenomenon is an illustration of what is sometimes known as Ostwald's step rule or law of stages.

The influence of the nucleation barrier is dramatically illustrated in the explosive decomposition of some metal azides. These unstable compounds (MN_3) decompose thermally to produce metal and nitrogen gas. When such a compound is heated, at first little reaction occurs. The compound remains stable until a small number of nuclei of metal form. Once nuclei are formed, the reaction may then proceed rapidly with gas evolution, in some cases explosively.

Growth

Once a nucleus is formed, the second stage of crystal formation involves the addition of orderly layers to produce the larger unit. The problem of the mechanism and velocity of crystal growth has been the subject of much research in recent years. In some cases it was found, for example, that the observed rate of growth was 10^{1300} (a very large number indeed) times faster than expected. As Kittel has put it, "an all-time record for disagreement between observation and theory."

A little reflection indicates that the problem of growth is similar to the problem of nucleation. Consider the addition of a unit block to the planar surface of a sodium chloride crystal (Fig. 13-3). The

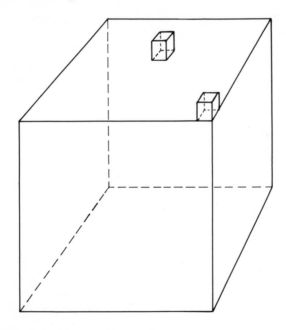

Fig. 13-3 Blocks added to a growing NaCl crystal. These blocks are almost as unstable as the nucleus of Fig. 13-1.

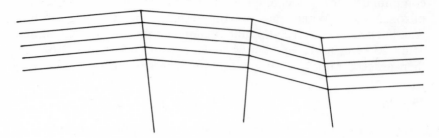

Fig. 13-4   A crystal surface showing slightly disoriented blocks.

four atoms which are added to form these blocks have only half their
valence forces satisfied, and thus the added block has a large energy,
and a large degree of supersaturation should be required to stabilize
it.   But it is observed that growth will occur quite rapidly at low
supersaturations, so the simple model of addition of blocks cannot
be correct.   This problem has been partly resolved by the discovery
of another type of imperfection in crystals, termed *dislocations*.

When x-rays are scattered from crystals at low glancing angles,
the crystal surface appears as a mosaic of small blocks, each block
with a more or less perfect structure, the blocks being slightly
disoriented with regard to each other as shown in Fig. 13-4.

The nature of dislocations and the manner in which these may
give rise to mosaic blocks with tilted grain boundaries are easily
pictured from Fig. 13-5, which shows the nature of a so-called edge

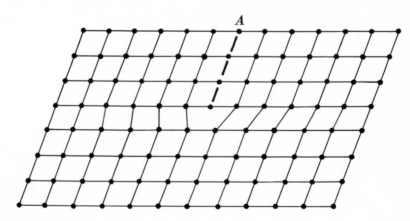

Fig. 13-5   An edge dislocation.   An extra line of atoms has been introduced
along line *A*.

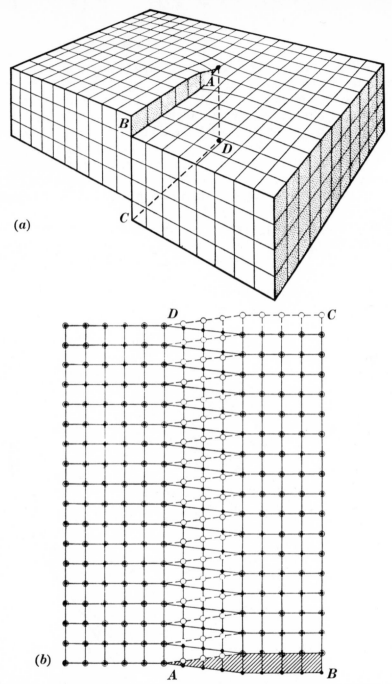

Fig. 13-6 (*a*) Appearance of a crystal with a screw dislocation originating along the line *AD*. (*b*) Arrangement of atoms in the plane *ABCD*. The black dots and circles represent atoms on either side of the dislocation plane. (*Adapted by permission from* "*Dislocations in Crystals,*" *by W. T. Read, Jr. Copyright* 1953, *McGraw-Hill Book Company, Inc.*)

dislocation.   In Fig. 13-6 is shown the screw dislocation, which may frequently be the controlling influence in crystal growth.   It is not difficult to see why such a dislocation should be important in growth. If a step is already present, an added unit introduces little extra surface energy, particularly when the unit is added near the end of the dislocation (Fig. 13-7).   A dislocation will tend to grow fastest at the end where valence forces are most satisfied, and hence as growth proceeds, the screw dislocation will be preserved and spiral around the origin.   A crystal surface as in Fig. 13-8 is convincing proof of the reality of this model.

Many mechanical properties of crystals are dependent on the number and type of dislocations present, for these represent regions of weakness.   Most real crystals are much weaker than theoretically anticipated.   Recently crystals of extreme perfection in the form of "whiskers" of diameter around 10^{-4} cm have been produced with strength approaching the theoretical limit.   It is beyond the scope of this book to discuss the complex features and importance of dislocations, but additional references will be found in the general bibliography on page 193.

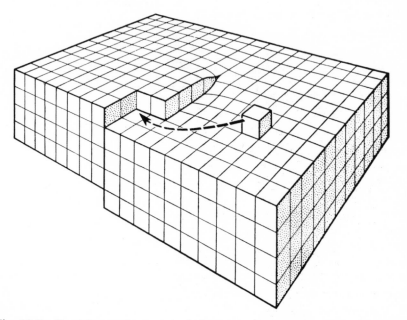

Fig. 13-7 Crystal growth on an existing screw dislocation. The dislocation must be preserved by block addition, but little extra surface energy is produced as would be the case in Fig. 13-3. (*Adapted by permission from "Dislocations in Crystals," by W. T. Read, Jr. Copyright* 1953, *McGraw-Hill Book Company, Inc.*)

Fig. 13-8 Growth patterns on silicon carbide caused by growth on dislocations. (*Adapted by permission from "Dislocations in Crystals," by W. T. Read, Jr. Copyright 1953, McGraw-Hill Book Company, Inc.*)

Crystal Shape or Habit

From our earlier discussion (page 53) of surface energy, it is clear that certain shapes of crystals should be energetically preferred. The surface-energy terms are not large, however, and some tolerance of shape is found. When crystals grow slowly, faces present are simple and are those that involve the production of planes in which the greatest number of valence forces are satisfied. When crystal growth is rapid—that is, under conditions of high supersaturation— and when the chances of recrystallization are slight, accentuated crystal habits may be found. No better example is found than in the magnificent dendritic forms of snow crystals.

If a crystal is grown in the presence of some other dissolved material, the habit is frequently modified compared with that produced in the absence of this material. Common modifiers for inorganic crystals are large organic molecules. For example, when sodium chloride crystallizes from pure water, simple cubes are normally formed. But if urea (NH_2CONH_2) is present in sufficient concentration, the salt crystallizes in octahedra. In any such case, if the organic molecule is absorbed on the surface, because of different atomic spacings on different possible planes, the molecule will be absorbed more strongly or specifically on preferred faces. Thus relative growth rates of faces may be modified and a change in habit produced.

Chapter 14

CHEMICAL REACTIONS INVOLVING SOLIDS

In this chapter we shall outline very briefly some of the more important reactions involving solids. This is a vast field described in an equally vast literature, but in the limited final bibliography the reader will find sufficient guides to more detailed discussions.

Reactions in Solids

Reactions in and between solids, which do not occur via a gas or solution phase, at some stage will involve diffusion of some species in the solid or across the surface of the solid. We have already discussed in Chap. 12 the nature of some simple defects in crystals and have indicated that diffusion processes, including electric conductivity, involve these rather mobile defects. To illustrate a simple example of diffusion, consider the self-diffusion in potassium chloride. Imagine that we place two cubes of KCl together as in Fig. 14-1a, with one cube being made of KCl containing the stable isotope K^{39}, and the other cube the β-active isotope K^{40}. We can follow the mixing of the activity in the two blocks by following the distribution of radioactivity. It would be found in this case that after different times the concentration of K^{40} through the crystal would appear as indicated in Fig. 14-1b. Further, the time to achieve a certain gradient would be an exponential (or logarithmic) function of the absolute temperature. These relations can be expressed in terms of some form of Fick's equation for diffusion:

$$P = -D \frac{\partial c}{\partial x}$$

This equation expresses the relation between P, the rate of permeation or the number of particles crossing unit cross section in the x direction in unit time, in terms of a constant D, the diffusion-coefficient characteristic of the solid and the given process in the solid, and the

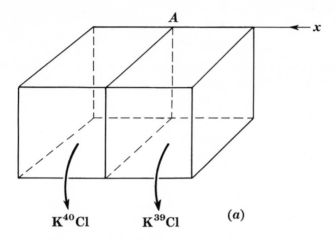

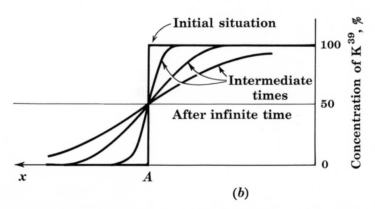

Fig. 14-1 (a) Two identical blocks, one of $K^{40}Cl$ and one of $K^{39}Cl$, are placed together. The isotopes mix with time, and the resulting concentration gradients are shown in (b) for various increasing times.

concentration gradient $\partial c/\partial x$, in the x direction. It will be noticed that in the direction in which diffusion occurs, the gradient must be negative. The diffusion coefficient is only constant at constant temperature and pressure and varies according to an equation of the form

$$D = D_0 e^{-E/kT}$$

or a plot of log D against $1/T$ should be linear. In this case E is termed the energy of activation of diffusion and represents the energy required to move the diffusing species in its defect position. In our example, if we could measure the conductivity of a KCl crystal due to potassium ions, we would find that this follows a similar relation:

$$\sigma = \sigma_0 e^{-E/kT}$$

Clearly the two processes involve similar things, and one might expect a simple relation between them. In this case, it can be shown that

$$\frac{\sigma}{D} = \frac{\sigma_0}{D_0} = \frac{Ne^2}{kT}$$

where N = Avogadro's number
$\quad\ k$ = Boltzmann constant
$\quad\ e$ = electronic charge

Such a diffusion process can thus be studied either directly or from a study of defect electric conductivity. Let us note carefully that in many ionic crystals the E terms are large, and thus such solid diffusion becomes rapid only at high temperatures, often approaching the melting point of the compound. A typical instructive and simple method of studying a solid diffusion process is illustrated in Fig. 14-2. The common mechanisms by which internal self-diffusion may occur are indicated in Fig. 14-3.

While a simple self-diffusion process is instructive in understanding the mechanism of such changes, more chemical and mineralogical

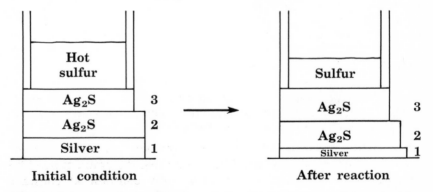

Fig. 14-2 Setup to show mechanism of the reaction between silver and sulfur. As reaction proceeds, silver block 1 diminishes, Ag_2S block 2 is unchanged, while Ag_2S block 3 grows. Only silver is diffusing in the Ag_2S layers.

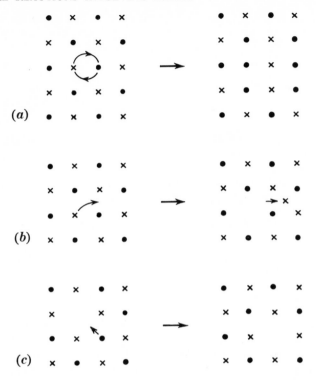

Fig. 14-3 The three common diffusion mechanisms: (*a*) rotational interchange; (*b*) interstitial diffusion; (*c*) vacancy diffusion.

significance must be attached to the reaction of two dissimilar solids in contact. The simple case of the reaction between two slabs of silver iodide and cuprous iodide is instructive. The most mobile species are interstitial Cu^+ and Ag^+ ions. In the simple mixing which occurs, one can imagine that the iodide anion framework remains fixed while mixing of the cations occurs. Clearly such mixing must be sympathetic, for neutrality requires that for every silver ion which crosses the boundary (Fig. 14-4), a cuprous ion crosses in the opposite direction.

Another type of process is illustrated when slabs of silver iodide and mercuric iodide are placed in contact. In this case a compound, Ag_2HgI_4, forms at the boundary. Once such a boundary layer compound is formed, it can grow only by diffusion, in this case diffusion of Ag^+ and Hg^{++} ions through the layer. Again the iodide framework is essentially fixed, only rearranging over short displacements where the compound forms. If a compound forms,

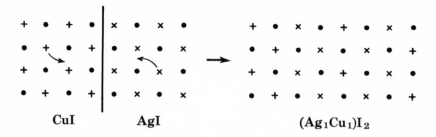

Fig. 14-4 Diffusion producing a homogeneous solid solution of AgI in CuI. The anion iodide lattice can be considered to remain stationary, while Cu^+ and Ag^+ ions diffuse via the interstitial mechanism.

as well as the diffusion barrier which will control the rate of advance of the layer of compound, an initial nucleation barrier must be crossed, and this frequently introduces what is known as an *induction period*, during which no reaction occurs.

The fixed anion model probably applies to many complex cases where the anions are very large relative to cations. For example, when slabs of MgO and Al_2O_3 are placed in contact, spinel $MgAl_2O_4$ forms at the boundary. Again diffusion of $3Mg^{++}$ and $2Al^{3+}$ across the layer could explain the process. If MgO and SiO_2 are placed in contact (Fig. 14-5), a layer of Mg_2SiO_4 (forsterite) will form on the MgO, and $MgSiO_3$ (enstatite) on the SiO_2. In this case the exact mechanism of the process may be complex, for the movement of a small highly charged cation such as Si^{4+} may be quite difficult compared with the movement of Mg^{++} and O^{--}.

One of the most simple of all solid-solid reactions is the process of recrystallization. As already stressed (page 172), an aggregate of small crystals is unstable with respect to larger crystals, and a crystal with an accentuated habit will be unstable with respect to some form more equidimensional because of surface-energy terms. If the grains are separated, transport will involve a solution or gas phase, and as the vapor pressures of most ionic crystals are extremely small near room temperature, fine particles in a dry atmosphere may survive for a long time—for our purposes, infinite time. Any medium in which the crystals have appreciable solubility will accelerate the process as the fine crystals will have a larger solubility than the coarse.

If efficient recrystallization of most solids in an inert atmosphere is to occur, the grain boundaries must be moved into intimate contact. A typical geological example of such a process is the recrystallization of a fine-grained limestone into a coarse marble by the influence of pressure-compaction and moderate heating. The driving force of recrystallization falls off rapidly with increasing grain size, but it is noteworthy that given geological time, survival of cryptocrystalline materials is uncommon.

All types of polymorphic changes involve limited movement of atoms, while with first-order transformations this is coupled with nucleation. An excellent example of such processes involves the reactions which occur on cooling a high-temperature solid solution of $NaAlSi_3O_8$ and $KAlSi_3O_8$. As we have mentioned previously (page 136), at elevated temperatures both albite and orthoclase occur in a disordered structure in which Al and Si atoms are randomly distributed and the two solids are completely miscible. When a 1:1 mixture is cooled, the mutual solubility becomes more and

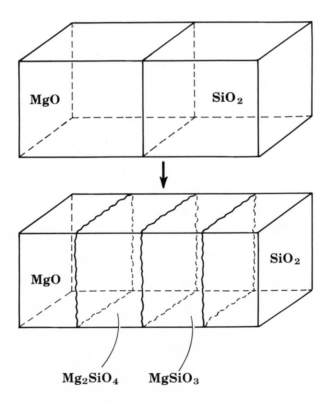

Fig. 14-5 Reaction of MgO and SiO_2. After a time the phase assemblages will form as shown, and gradually the MgO and SiO_2 will be reduced.

more limited, and at low temperatures the mixture, if it passes to its most stable configuration, will consist of nearly pure ordered albite and nearly pure ordered orthoclase.   Such an ideal situation is rarely achieved.   If the mixture is cooled fairly rapidly, some unmixing will occur, but the scale of the process may be so fine that only x-rays can detect it.   Further the feldspars will probably

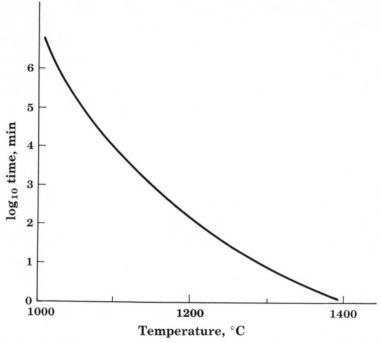

Fig. 14-6 Relation between the logarithm of reaction time and temperature for the solid-state reaction

Kyanite $\rightarrow$ mullite + cristobalite

remain disordered.   If the cooling is more gradual, the mixture will separate into visible patches of the two phases and more ordering will occur.   From such observation we may suggest that in this process the nucleation process is relatively facile, diffusion to grow the nucleus more difficult, and the diffusion involved in the order-disorder process most difficult.   To give a better idea of the difficulties in some solid-solid reactions, in Fig. 14-6 we have plotted the time for 1 % of the reaction

Kyanite $\rightarrow$ mullite   + cristobalite
$$3Al_2SiO_5 \rightarrow Al_6Si_2O_{13} + SiO_2$$

to occur as a function of temperature in the dry state.

Solid-Gas Reactions

In this section we shall consider some typical reactions of solids with gases at low pressures where the gas has negligible solvent activity; i.e., the gas phase contains only volatile constituents.

The simplest of such processes is that of sublimation (the reverse, condensation, is more difficult in that nucleation must occur). Typical examples are:

$$ice \rightarrow molecules\ of\ H_2O$$
$$NH_4Cl_{solid} \rightarrow NH_{3gas} + HCl_{gas}$$
$$S_{solid} \rightarrow S_2\ etc.\ gas$$
$$HgS_{solid} \rightarrow Hg_{vapor} + S_{vapor}$$

While sublimation is most uncommon with ionic crystals and minerals, except at extreme temperatures, nevertheless it must be remembered that all solids do have a finite vapor pressure or do pass to some extent into the gas phase. Thus the pressure of HgI_2 gas above the solid is 10^{-5} atm at 65°C, a pressure indicating an appreciable number of atoms in the gas phase; with potassium chloride at 570°C, the vapor pressure is 10^{-6} atm; but at 25°C, only 10^{-30} atm. To rapidly transport most solids in a vapor, the temperature must be raised well toward the melting point of the solid.

While all solids have a finite vapor pressure, many dissociate into another solid and a vapor. Such a process could be termed incongruent vaporization to emphasize its similarity to incongruent solution and melting. In all such cases, the solid and vapor phases have different compositions or ratios of the constituent atoms of the solid. Typical of such reactions are the dissociation of oxides, hydrates, and carbonates.

$$M \cdot (OH)_{2solid} \rightarrow MO_{solid} + H_2O_{vapor}$$
$$M \cdot H_2O_{solid} \rightarrow M_{solid} + H_2O_{vapor}$$
$$MO_{solid} \rightarrow M_{solid} + O_{vapor}$$
$$MCO_{3solid} \rightarrow MO + CO_{2vapor}$$

These processes again occur via steps of nucleation, diffusion, and growth, and any of these steps may control the rate of the process. Frequently, nucleation is the difficult step, so that the compound can be substantially superheated before it eventually produces nuclei to allow the process to proceed.

The vapor pressure over such compounds, and also over congruently subliming compounds, is found to vary exponentially with the absolute temperature at low pressures.   The slope of the vapor pressure-temperature curve depends partly on the energy required to break the material apart into vapor or solid and vapor.   Typical vapor pressure curves for a salt and a hydrate are shown in Fig. 14-7.

One of the most interesting types of reaction shown by solids in the presence of gases is the so-called *tarnishing reactions*.   These tend to fall into two classes: reactions which tend to completion, and those where a protective film is formed and further reaction is inhibited.   Were it not for this latter effect with materials such as aluminum and stainless steel, man's concern with corrosion would be greater than it already is.   Typical of the completion reaction is the tarnishing of metallic silver by sulfur vapor.   In this case (see Fig. 14-2), a layer of Ag_2S forms on the silver, and then further growth of this layer is controlled by the rapid diffusion of silver ions and electrons through the sulfide layer to meet the sulfur vapor condensing on the surface.   In some cases where a metal has several oxidation states, the layers may be complex, e.g., the reaction of iron with oxygen, as shown in Fig. 14-8.

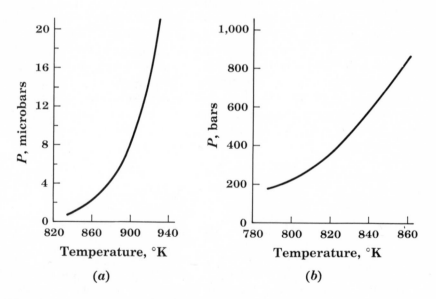

Fig. 14-7   (*a*) Vapor pressure of KCl molecules over crystalline KCl;   (*b*) vapor pressure of water over crystalline $Mg(OH)_2$. The unit of pressure in (*a*) is millionths of a bar, and in (*b*) bars. Note that 1 bar = 0.987 atm.

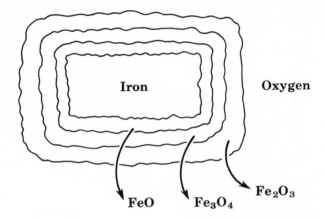

Fig. 14-8 Possible zonation during the partial oxidation of iron.

When a metal such as aluminum is placed in oxygen, a layer of oxide about 40 A thick is formed on the surface, and then reaction virtually ceases. In this case it is not too difficult to see why this occurs. The growth of the film depends on the migration of Al^{3+} ions and electrons through the Al_2O_3 layer. Once the layer forms, oxygen atoms condense on the surface; these combine with electrons and form O^{2-} ions. Thus a negative potential is established on the surface layer and exerts a potential drag which assists in pulling the Al^{3+} ions through the layer. In this case the diffusion of aluminum is difficult, and as the film grows in thickness, the assistance from the electrostatic pull becomes smaller, and eventually the process ceases to operate at low temperatures.

Solid-solution Reactions

In any solvent or liquid a pure solid has a definite solubility. The solubility may be congruent:

$$AgCl_{solid} + H_2O_{liquid} \rightarrow Ag^+_{soln} + Cl^-_{soln}$$

$$SiO_{2_{quartz}} + H_2O_{liquid} \rightarrow Si(OH)_{4_{soln}}$$

and involve the formation of ions or neutral molecules, or may be incongruent:

$$K_2PbCl_{4_{solid}} + H_2O_{liquid} \rightarrow 2K^+_{soln} + 2Cl^-_{soln} + PbCl_{2_{solid}}$$

In all these cases we may say that in the saturated solutions at a given pressure and temperature,

$$\text{concn of Ag}^+ \times \text{concn of Cl}^- = \text{const}$$

$$\text{concn of Si(OH)}_4 = \text{const} \qquad \text{etc.}$$

If more than one polymorphic modification of a solid is possible, the most stable form will have the lowest solubility. Perhaps the great majority of sedimentary and metamorphic processes in geology proceed via congruent or incongruent solubility in an aqueous solution. Incongruent solution may lead to leaching. Recrystallization processes are also promoted by a solvent, for again a solution in equilibrium with a fine material is supersaturated with respect to coarse material. In many solid reactions, the presence of a good solvent promotes reaction for it allows the bypassing of a difficult diffusion path.

The theoretical calculation of the solubility of a solid in a liquid is still generally impossible, but the factors involved are well known. Ionic crystals tend to be soluble only in liquids formed from molecules with large dipole moments. Such liquids normally also have high dielectric constants. The strong forces between ions in a lattice can only be broken, and solution can only occur, if equally strong ion-ion or ion-dipole forces operate in the solution. Covalent solids producing molecules in solution tend to dissolve in solvents of low polarity and are insoluble in solvents such as water. In this case the forces between the water molecules exceed the forces between a water molecule and the covalent molecule, and thus the solute species cannot squeeze into the solvent structure. But over-generalization is dangerous. In any actual case one must assess the balance between:

1. Forces binding the solid
2. Forces between the liquid molecules
3. Forces between solvent and solute molecules

Consider the solution of quartz in water. Quartz (SiO_2) can probably be considered dominantly a covalent solid. The Si—O bonds are very strong. In water the hydrogen bonds holding the liquid together are moderately strong. When SiO_2 dissolves, present models suggest that the reaction

$$SiO_2 + 2H_2O \rightarrow Si(OH)_4$$

occurs, and hence the four strong Si—O bonds in the solid are retained in the dissolved state. Further, the four strong O—H bonds in the two water molecules are also retained. Finally, it is clear that the $Si(OH)_4$ molecule may form hydrogen bonds with water molecules in the liquid and should mix with the liquid. All facts considered, we would expect quartz to have appreciable solubility in water.

As already mentioned, crystallization and recrystallization are promoted by a solvent. Depending on the nature and impurities in the solvent, different planes of a crystal grow at different rates. Solution also occurs on preferred lattice sites and leads to etching of crystals. As noted (page 173), crystal growth may frequently be controlled by dislocations. In the solution process, dislocations again are favored positions of attack, and where dislocations meet the surface, etch pits tend to form and may be used to estimate the number and position of these defects. In mineralogy, etch pits have additional significance, for the shape of an etch pit may indicate the true symmetry of a crystal lattice which may not be shown by the external form of the crystal.

Adsorption-Absorption and Catalysis

Adsorption occurs when some foreign species is fixed to the surface of a solid. Absorption involves the preliminary surface adsorption followed by diffusion into the interior of the solid. Molecules adsorbed on a surface are frequently deformed so that a chemical reaction will occur on the surface which is difficult to perform in the absence of the surface. Such an effect, much used in all of preparative chemistry, can be termed *surface catalysis.*

The forces causing surface adsorption are commonly van der Waals, ion-dipole (for example, H_2O on NaCl), hydrogen bond (for example, H_2O on a clay or hydroxide), or normal chemical bonds. Thus when a water molecule is adsorbed on a metal oxide or on quartz, hydroxyl ions may form. Such a process is sometimes termed *chemi-sorption.* If selective adsorption is to occur, it is necessary that some periodicity in the solid lattice match the charge distribution in the adsorbed molecule, and it is not difficult to see that by shrewd choice of surface, a molecule may be deformed into reactive or unreactive configurations. Many gas-solid, liquid-solid reactions involve a preliminary adsorption or desorption step, and the initial lattice may determine the product lattice (see page 171).

Absorption processes are well demonstrated by the swelling of clays when water molecules penetrate the silicate-hydroxy sheets

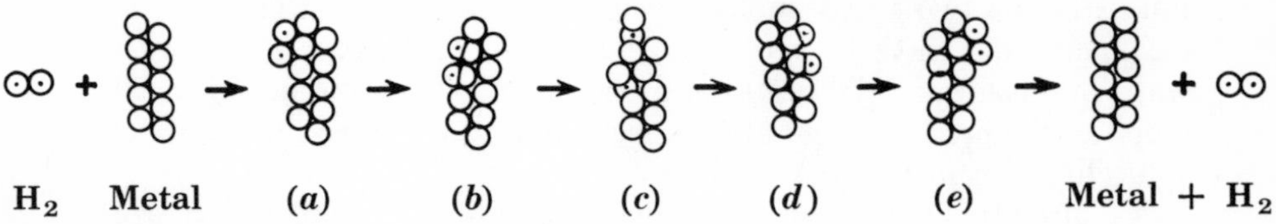

H_2 Metal   (a)   (b)   (c)   (d)   (e)   Metal + H_2

Fig. 14-9 Possible stages in the passage of a hydrogen molecule through a metal membrane: (a) hydrogen molecule is adsorbed on the surface; (b) hydrogen molecule is dissociated and the hydrogen atoms "chemi-sorbed"; (c) hydrogen atoms diffuse into the metal, disrupting the original structure and forming M—H—M bonds; (d) hydrogen atoms appear on the far side of the membrane, "chemi-sorbed"; (e) atoms recombine and finally diffuse away.

forming hydrogen bonds, and with de-gassed zeolites (see page 128) where molecules such as water, iodine, and hexane may be absorbed in the open channels.

A classical case of sorption is shown by the ability of metals such as palladium to accept large quantities of hydrogen gas. Such metals can serve as excellent semipermeable membranes (see Fig. 14-9). In these cases, however, the hydrogen molecule is drastically deformed in its passage through the metal. The entire process of passage probably proceeds by (a) adsorption of H_2 molecules on surface, (b) surface dissociation of H_2 into 2H atoms, (c) formation of hydrogen-metal bonds, (d) recombination of H atoms, and (e) desorption.

The Photographic Process

There is probably no process that involves a combination of all types of solid reactions so dramatically as the photographic process. A photographic plate consists of fine-grained crystals of silver bromide set in some suitable emulsion. When these grains are exposed to a limited number of light quanta, a reaction occurs which allows chemical development of the so-called invisible latent image to produce our negative. Let us examine the proposed model.

When the AgBr grain absorbs the visible photon, a process

$$Br^- + h\nu \rightarrow Br + e$$

occurs. Eventually the positive hole created in the bromine reaches the surface (see Fig. 14-10) and produces a free halogen which is adsorbed on surface. Next an interstitial silver ion (there are plenty in AgBr) captures the liberated electron and becomes a silver atom.

A second photon produces a second hole and free bromine so that a Br_2 molecule is formed on the surface and escapes. The electron freed is captured by the neutral silver to form Ag^- which in turn captures an additional interstitial ion to become Ag_2.

A permanent memory of the photons absorbed is now present in the crystal. If overexposed, this latent image will grow by this mechanism till visible silver specks appear. If not, only those grains containing the latent image can be developed by placing the crystal in a suitable reducing agent. This developer, in solution, passes electrons to the latent invisible silver specks and charges them negatively so that they now trap more interstitial ions which, remarkably enough, diffuse through the solid, not the liquid, phase. By adjusting the reduction potential of the developer, only those grains with a latent image are developed. The development again produces visible grains of silver on the latent grains. Thus our process of exposure and development involves

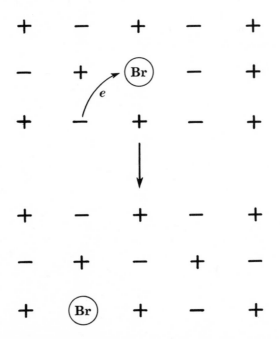

Fig. 14-10 Migration of a "positive hole." In this case the positive hole is at the site of a neutral bromine atom. By electron exchange, the positive hole or neutral bromine appears at the surface.

1. Production of positive holes and electrons
2. Diffusion of positive holes
3. Adsorption of atoms
4. Diffusion of interstitial ions
5. Desorption of molecules to the gas phase
6. Nucleation of silver
7. Electron transfer from solution → solid

altogether a remarkable example of the reactivity of the defect solid.

FURTHER READING

Chemical Bonding

Coulson, C. A.: "Valence," 2d ed., Oxford University Press, London, 1961.
Pauling, L.: "The Nature of the Chemical Bond," 3d ed., Cornell University Press, Ithaca, N.Y., 1960.

Crystal-field Theory

Orgel, L. E.: "An Introduction to Transition-metal Chemistry," Methuen & Co., Ltd., London, 1960.

Solid State, Defects, Diffusion, and Dislocations

Barrer, R. M.: "Diffusion in and through Solids," Cambridge University Press, London, 1951.
Jost, W.: "Diffusion in Solids, Liquids, Gases," Academic Press Inc., New York, 1952.
Kittel, C.: "Introduction to Solid State Physics," 2d ed., John Wiley & Sons, Inc., New York, 1956.
Read, W. T., Jr.: "Dislocations in Crystals," McGraw-Hill Book Company, Inc., New York, 1953.

Structural Chemistry and Mineralogy

Berry, L. G., and Brian Mason: "Mineralogy," W. H. Freeman and Company, San Francisco, Calif., 1959.
Bragg, W. L.: "Atomic Structure of Minerals," Cornell University Press, Ithaca, N.Y., 1937.
Deer, W. A., R. A. Howie, and J. Zussman: "Rock Forming Minerals," Longmans, Green & Co., Ltd., London, 1962.
Hurlbut, C. S.: "Dana's Manual of Mineralogy," 17th ed., John Wiley & Sons, Inc., New York, 1959.
Wells, A. F.: "Structural Inorganic Chemistry," 3d ed., Oxford University Press, London, 1962.

Crystallization

Vanhook, A.: "Crystallization Theory and Practice," A.C.S. Monograph 152, Reinhold Publishing Corporation, New York, 1961.

Thermodynamics—Chemical Stability

Denhigh, K. G.: "The Principles of Chemical Equilibrium," Cambridge University Press, London, 1957.

Guggenheim, E. A.: "Thermodynamics," 2d ed., North Holland Publishing Company, Amsterdam, 1950.

Lewis, G. N., and M. Randall: "Thermodynamics," 2d ed., revised by K. S. Pitzer and L. Brewer, McGraw-Hill Book Company, Inc., New York 1961.

INDEX

The atomic weight of an element is shown above

1.008 H 1								
6.940 Li 3	9.013 Be 4							
22.990 Na 11	24.32 Mg 12							
39.100 K 19	40.08 Ca 20	44.96 Sc 21	47.90 Ti 22	50.95 V 23	52.01 Cr 24	54.94 Mn 25	55.85 Fe 26	58.94 Co 27
85.48 Rb 37	87.62 Sr 38	88.92 Y 39	91.22 Zr 40	92.91 Nb 41	95.95 Mo 42	98 Tc 43	101.1 Ru 44	102.91 Rh 45
132.91 Cs 55	137.36 Ba 56	Rare earths	178.6 Hf 72	180.95 Ta 73	183.85 W 74	186.22 Re 75	190.2 Os 76	192.2 Ir 77
223 Fr 87	226.05 Ra 88	Actinide series						

	138.92 La 57	140.13 Ce 58	140.92 Pr 59	144.24 Nd 60	145 Pm 61	150.35 Sm 62	152.0 Eu 63
Rare earths							
Actinide series	227 Ac 89	232.05 Th 90	231.1 Pa 91	238.07 U 92	237 Np 93	242 Pu 94	243 Am 95